GENETIC SPECIES AND ECOSYSTEM DIVERSITY

Encyclopaedia of Environmental Pollution and Awareness in 21st Century Series

GENETIC SPECIES AND ECOSYSTEM DIVERSITY

Edited by
V.K. Prabhakar

ANMOL PUBLICATIONS PVT. LTD.
NEW DELHI - 110 002 (INDIA)

ANMOL PUBLICATIONS PVT. LTD.
4374/4B, Ansari Road, Daryaganj
New Delhi - 110 002

Genetic Species and Ecosystem Diversity

First Edition, 2001

ISBN 81-261-0904-1

PRINTED IN INDIA

Published by J.L. Kumar for Anmol Publications Pvt. Ltd., New Delhi - 110 002 and Printed at Mehra Offset Press, Delhi.

Contents

PREFACE

Today, the cry of 'pollution' is heard from all the nooks and corners of the world, and pollution has become a major threat to the very existence of mankind on this earth. It is the major challenge of our times.

The pollution of various resources has gone to such an extent that we are unable to breathe fresh air and drink fresh water. On one hand the advancements of science and technology have added to the human comforts by giving us automobiles, electrical appliances, supersonic jets, space crafts, better medicines, better chemicals to control harmful insects and other pests, etc., but, on the other hand, they have given us a very serious problem to face that is pollution. Pollution can be defined as an undesirable change in the physical, chemical, or biological characteristics of the air, water, or land that can harmfully affect health, survival, or activities of humans or other living organisms.

Pollution does not have to cause physical harm, but it may merely interfere with human activities. Chemicals spewed into the air or water from an industrial plant may be harmful to humans and other organisms living near the plant. If expensive pollution controls are required, the plant may be forced to shut down.

India faces serious problems of environmental pollution. Water pollution from untreated community and industrial wastes flowing into our rivers and streams, increasing use of fertilizers and pesticides, dumping of organic wastes etc., are issues that need to be tackled on the priority. Air pollution, long regarded as the bane of industrialised nations looms large over our country.

Here, we are presenting this *Encyclopaedia of Environmental Pollution and Awareness in Twenty-first Century* in fifty volumes with a view to make people conscious about the seriousness of pollution in the present time and its remedies.

Editor

1 BIODIVERSITY

The concept of the balance of nature gained credence in the early 18th century when Isaac Newton introduced very successfully the notion that nature could be explained in terms of a few simple laws. According to the Newtonian view, the world is formed by elements that are simple and that respond to regular and deterministic dynamics.

Today, physicists believe that nature is complex, not simple, and that the Newtonian view is insufficient as a general explanation of how the universe functions. The universe, physicists are finding, is not in balance.

Disturbances and irregularities of all sorts are no longer seen as aberrations but as integral parts of nature. Physicists are also discovering that at the very origin of the cosmos, at the so-called big bang, a singular, irreversible, and complex universe was produced. Similarly, ecologists have been observing and documenting that most, if not all, ecosystems are not in balance. But why should scientists be concerned with whether an ecosystem is in equilibrium? The answer to this question is crucial for understanding biodiversity. Systems not in equilibrium behave very differently than do systems in equilibrium. Their behaviour can even appear strange and mysterious.

For example, when the source of a disturbance is removed from a system that is near equilibrium, that system is expected to return to its previous state. When, in an undisturbed forest, a gap is created by landslide or a falling tree, new seedlings grow in the gap and restore the equilibrium.

After a few years, it is almost impossible to tell that there once was a gap. But when the source of change is removed from systems that are not in equilibrium, they do not return to equilibrium. Instead, they adopt a new state. For example, when land used for agriculture in the Amazon forest is abandoned, it grows into a grassland or a savanna and does not return to the original forest.

Natural disturbances are a necessary part of the forest ecosystem's function. For example without gaps creating disturbances, most forests cannot renew themselves. And fires, storms, and hurricanes are and have been a part of life since the beginning of the planet; without them, ecosystems could not function properly. Knowing about the features of non-equilibrium systems is important for proper ecosystem management.

Because a non-equilibrium ecosystem does not necessarily revert to its previous state when the convulsion is removed it is often impossible for humans to return ecosystems to their original condition. Therefore, one set of management guidelines will not suffice for all situations. The previous history of changes and the present disturbance regime will determine the consequences of any management.

There are two conditions that cause population diversity. First, new genotypes are constantly cropping up in a population through mutation, recombination, and related genetic phenomena and through immigration of individuals, their gametes, or their propagules. Second, diversity in the population is eliminated by natural selection and lost through emigration of individuals. Every genetic variation, from gene mutations to entire species, will disappear eventually.

This loss can be a very fast process, or the variants can survive for a long time. Species that have survived for extended periods include horseshoe crabs, which have been around for 200 million years, and cockroaches, which originated even earlier, in the Carboniferous period. The speed at which new variations originate in relation to the rate at which they are eliminate determines the actual diversity of the system.

In the process of gene mutation, all heritable diversity ultimately arises at the molecular level. Gene mutations are chemical changes that take place in the composition of the DNA (deoxyribonucleic acid) molecule, the chemical substance responsible for heredity DNA is remarkable in that it shows complex chemical behaviour not expected from systems in equilibrium. The most remarkable characteristic of DNA is its ability to regenerate itself, its "autocatalytic" behaviour.

The DNA molecule consists of four kinds of repeating units called uncleotides, which can be arranged in any sequential order. DNA has the curious ability to maintain its physicochemical integrity regardless of the order of the four nucleotides. That is, any DNA molecule behaves like any other of the same length, regardless of the nucleotide content.

This characteristic of DNA makes life feasible because if only one arrangement were possible, or if the chemical stability of one arrangement were significantly different from that others, then all DNA molecules would be alike. The order of the nucleotides uniquely determines the characteristics of the chemical products made by DNA. But the order of the nucleotides in a DNA molecule can change, and these changes are called mutations. The regular appearance of *mutations* gives life its diversity.

Mutated molecules of DNA reproduce their changes and make *modified enzymes,* which, in turn, make *altered cells.* Mutated cells result in *modified organisms.* Thus, new mutations permit the evolution of new characteristics by changing the *structure* and /or *function* of enzymes and other *proteins.*

The characteristics of the enzymes and proteins are determined by those of the DNA molecule, but not the reverse — the "information" flows only in one direction. This flow contrasts with that of an ordinary chemical reaction, in which there are forward and backward reactions. The appearance of mutations is a *random* process. There is no way to anticipate or predict what mutation is going to occur, only that some will. It is natural *selection,* however, that determines whether new mutations get established or are eliminated. The processes of mutation and selection can be looked at in two ways.

Mutation and selection might be random, independent processes that are not influenced by the characteristics of the system, in which case no active process of diversity maintenance exists. On the other hand, a certain degree of variety may be required for living systems to function properly. If so, there may be *system feedbacks* that modify the rates of mutation and selection. Below a certain threshold, a system may collapse because of insufficient variety. According to this last view, diversity is actively maintained in a system.

Mutation and natural selection generate genetic and morphological variation within a lineage. Yet what is most distinctive in a community is the coexistence of different species, each acting independently and not sharing its DNA or mutations. How, then, does the great variety of species originate? Speciation is the process that separates genetic variation into distinct units, or species. In speciation, the original population of organisms with similar genes, called a gene pool, is divided into two or more gene pools.

Each of these new gene pools then acquires a unique set of characteristics (cellular, tissue, organ, and organismic) through mutation and selection. (The mechanisms that create species have been a focal point of research in the field of organic evolution.)

Speciation follows most commonly from the physical division of a gene pool. The separation inhibits interbreeding between individuals in the two populations. In time, one or both populations change enough to prohibit interbreeding.

Because the fate of every species is to become extinct eventually, there must be an influx of new species for life to continue. If, presently, there are about 10 million species, and, on average, each species has a life span of 1 million years, then an average of 10 new species must originate each year.

On the other hand, if the life span of a species is only 100,000 years (as many biologists argue), then 100 new species must originate each year. Scientists are often aware of a species' extinction, but the appearance of new species is difficult to observe. Because most species are insects and other small invertebrates, most new species fall into these two poorly studied groups. Thus, there are few well-documented cases of speciation.

Certain individuals in a population survive and reproduce, while others die without leaving offspring. To understand this phenomenon, scientists must first realize that there usually is not enough of some environmental resources (food, water, minerals) to go around. The availability of resources in the environment limits the rate at which individual organisms can reproduce and their ability to survive.

Nucleic acid replication qnd cell, tissue, and organism growth and reproduction all require energy and materials, of which there are only limited amounts. Individuals that are efficient at harvesting resources have a higher probability of reproducing and surviving than do those that are inefficient.

Mutations in DNA provide species with a constant input of physical and chemical variations, some of which improve the efficiency with which the species selects and harvests resources. New mutations may increase the efficiency of protein synthesis at the molecular level, or they may improve the water uptake by a plant at the organism level. Mutations may even increase the efficiency with which a mutualistic association between a plant and its specific pollinator develops at the community level.

DNA molecules that are dissimilar in the order of their nucleotides usually do not have characteristics that directly increase or decrease their own survival or reproductive capacity. Given the appropriate environment and resources, all DNA molecules reproduce at the same rate despite their nucleotide composition.

DNA molecules manifest their distinctive survival rates through their effects on the living organisms in which they are embedded. These organisms, be they single cells living in pure culture or multicellular organisms living in complex communities, will inevitable differ in their ability to garner resources.

Therefore, they will vary in their survival and reproductive capacities and, consequently, so will the DNA in their cells. Natural selection is an optimizing process. It is process that favours efficiency and produces adaptation. It is not a process that involves chance, like the appearance of mutations, but it must operate with the elements that it has.

Thus, natural selection never makes a "perfect" organism. Because the environment is constantly changing, a perfect organism, if it could be produced, would soon cease to be perfect as the environment changed. Such systems, in which the optimum state is favoured but cannot be attained or maintained, are called frustrated systems.

Diversity is the result of two opposite actions: the processes that produce new genotypes, new varieties, and new species and the processes that eliminate mutations, variants, and species from the system. Natural selection is primarily responsible for the reduction of biodiversity; it acts through differential reproduction and differential mortality.

In order words, the probability of reproducing and the probability of dying are not the same for all organisms. Rather, reproduction and death are related to the organism's characteristics as well as to the environment in which that organism lives.

How is species diversity linked to ecosystem structure and function, to the different ways in which organisms and population interact with each other, and to those community properties that emerge from these interactions? Communities have structures and properties not possessed by the populations within them that are called emergent properties, which include trophic structure, stability, guild structure, and successional stages. Many scientists believe that species diversity is essential for the proper functioning of communities and for the emergence of community-level properties. Just as many different DNA-encoded enzymes are needed for a complex organism to function properly, so, scientists believe, are many kinds of species necessary to maintain community structure. But is any diversity sufficient, or are specific mixes of species necessary for the communities and ecosystems of function? This is a very old question in ecology, and two opposing views exist.

One view is that a community is formed by the species that happened to arrive first — that the mix of species in a community is a matter of chance: "The vegetation of an area is merely the resultant of two factors, the fluctuating and fortuitous immigration of plants, and an equally fluctuating and variable environment."

According to the opposing view, "In any fairly limited area, only a fraction of the forms that could theoretically do so actually from a community at any one time.... The community really is an organized community in that it has a 'limited membership.' "

Are these two very different views really irreconcilable? Not necessarily, if a community is defined as the sum of all the plants and animals that grow together in an area, certain patterns can be observed: Areas with more than 1,000 millimeters of evenly distributed rainfall always contain a woodland; conifers prevail in areas of extremely low winter temperatures; trees in warm climates have broad leaves; and succulent plants are found in dry climates. The same is true for animals.

The mix of species encountered in a community is not a random sample of all plants and animals in the world. Yet, under close scrutiny, differences that are difficult to explain can be found between similar localities. Ecologists cannot predict what species will occur in a given climate and soil type. Also, there have been cases where dominant species, such as the American chestnut, have been lost, with apparently little effect in the overall working of the biological community.

The community behaves in a "frustrated" fashion. Two types of processes are at work and neither is dominant. First, there is natural selection. Species in the community are constantly evolving to increase their ability to withstand the rigours of the environment. This process accounts for cold-hardy species near the popes, for grasseating ruminants in savannas, and for fruiteating bats in forests. The individuals that escape their predators and survive other environmental challenges pass on their characteristics to their offspring.

Second, there is chance. When the climate changes, characteristics that were once advantageous can become a burden. Diseases might be introduced, such as Dutch Elm disease, for which local species have no defense. Hurricanes, tornadoes, floods, and droughts are unpredictable and their frequency changes over time. Chance also plays a role in dispersal. For instances, a species might be absent from a community purely by chance. Thus, both natural selection and chance result in the steady coming and going of species through immigration, emigration, extinction, and gene mutation.

A large number of additional unanswered questions exist regarding the role of biodiversity in communities. For example, can more and more species be packed into a community, or are there upper and lower limits? A related question concerns optimal levels of diversity and the factors that control them. Other issues include the role of different individuals within a population, different populations within a species, and so on.

Do the members of the community collaborate in the efficient use of energy and resources? In a variable environment, the existence of individuals with different characteristics may increase the ways in which the population can respond to change.

For example, if all plants of a species had similar water requirements, all of them would suffer water stress in any year that was drier than normal, and such periods would result in significantly reduced seed production. But if there is genetic variation, some individuals might perform above average each year within certain limits of environmental variation.

Therefore, seed production would be satisfactory in both wet and dry years. Moreover, genetically adaptable organisms should survive in more variable environments than do genetically uniform populations. Experience with crops shows that highly productive but genetically uniform varieties have more restricted environmental requirements than do less productive but more variable varieties. Also, plantations formed by uniform varieties are more susceptible to pest and disease outbreaks.

So, variability within a species seems to be important for long-term survival. Does the same hold for a community, which is formed of species that live together but do not have a common gene pool? Communities with high species diversity may also cope with long-term environmental fluctuates better than do communities with few species.

However, the evidence is contradictory. Terrestrial communities in the climatically variable midlatitudes are less diverse than tropical communities in more uniform environments. Also, deep benthic communities are among the most variable communities anywhere though they exist in possibly the most equitable environment on the planet.

According to modern ecological theory, every species in a community occupies a singular ecological niche, or a role in the environment. A niche is thought of as a multidimensional space where each dimension is a characteristic of a species. Because every species has at least one physical or behavioural characteristic that separates it from other species, every species has a unique niche. In principle, niche theory supplies the theoretical framework to explain the number and types of species in a community. Niche theory predicts that communities that vary in total resources or in the quality of those resources will contain more niches and, therefore, more species.

Nevertheless, resources are sometimes partitioned differently in communities with very similar resources. In other words, communities may be have large or small niches, which result in a few "generalist" species (those with board niches) or many "specialist" species (those with small niches).

Finally, communities with similar resources and niche sizes may still vary in the number of species because of different degrees of niche overlap. Niche overlap is the degree to which two species exploit the same resource.

Communities differ in niche size and overlap for a variety of reasons, such as differences in climatic stability and predictability, spatial heterogeneity, primary productivity, computation predation, and degree of disturbance. Thus, niche theory is unable to predict the patterns of species diversity.

Using niche theory to understand why communities vary in number of species has proven very difficult in practice. Many studies have shown very elegantly how species partition various resources in space and in time, but it has proven very hard to show the precise contribution of each factor.

A key question is whether two species can have the same niche. In other words, is there species redundancy in a community? Present theory predicts that two species with identical niches cannot coexist, and empirical studies seem to verify this. Still, the question of how similar two coexistent species can be remains to be answered.

Additional unanswered questions concern the role of diversity in food webs. All organisms get their energy from the sun, either directly or indirectly. At the simplest level, cells of the green tissues of plants and some micro-organisms capture the sun's energy. Organisms also need specific chemicals from the environment to develop. For instance, all plants obtain the materials for development from soil or water and the air.

With these materials and energy from the sun, plants create energy-rich,complex chemical compounds, such as carbohydrates, cellulose,starch, proteins, fats, and wood. Animals ingest and extract the energy from these compounds and incorporate some of them into their bodies.

Decomposers reduce them to simple ions that eventually are returned to the soil or water. In a cyclic fashion, then, energy and chemical substances move from the environment to plants, to animals, to decomposing microorganisms, and back to the environment. This trophic diversity, or "food web", is essential for the proper operation of ecosystems.

Many studies have shown that the trophic structure of ecosystems is complex and varies from one ecosystem to another. Understanding how food webs function is essential for understanding the place of humans in the ecosystem,[19] because humans also derive their energy and materials from plants and animals.

Moreover, interpretation of food webs is necessary to understand the effect of toxic compounds released into the environment and the effect of the introduction of species into new areas. Are diverse ecosystems with complex food webs more stable than simple ecosystems? Stability, in this instance, refers to the ability of an ecosystem to withstand disturbances, including those produced by people. Scientists used to believe that simple systems are less stable than complex ones.

Clearly, there is a threshold of diversity below which most ecosystems cannot function. For example, no ecosystem can function without some plants and decomposers. Without plants, solar energy cannot be captured, and without decomposers, substances are not returned to the environment. But, is there a threshold below which present complex ecosystems lose their stability? How many different sorts of plants and animals are needed? and, if there are many of each sort, does the ecosystem become more stable? Productivity, or the total quantity of energy captured and biomass produced by an ecosystem in a year, is related to the issue of stability.

Are ecosystems that are formed by many species more productive than those with few species? It would appear that simple systems, both natural (*Spartina* marshes and high-latitude marine systems) and artificial (agricultural systems), are more productive than diverse ones. However, some scientists fear that these systems are not stable and have argued that society's increasing reliance on the productivity of these simple systems could eventually put humans in jeopardy.

Apart from the ethical and aesthetic reasons, is there any reason to fear that human survival is at risk if biodiversity is not preserved? Can humans exist surrounded only by agricultural fields, planted forests, and the like? This question is not easy to answer. In the short term, natural ecosystems can probably lose species without any great impairment of function. For instance, many planted forests are much simpler than the natural forests they replaced.

As long as the environment does not change very much, ecosystems can apparently lose many of their rare species without any visible effects. In some cases, common animal species, such as passenger pigeons, have disappeared or been drastically reduced without endangering human survival. Yet, environments inevitably changes, and sometimes drastically. It is then that the role of obscure species sometimes becomes very important.

When dinosaurs became extinct, an obscure group, primitive mammals, suddenly evolved into the dominant role they still play. And now, humans seem on the verge of a drastic changes, one of their own making—the result of global warming induced by increased carbon dioxide in the atmosphere.

Predicate climatic changes pose a series of specific questions relating to biodiversity for which there are presently no answers. What is the impact of elevated atmospheric concentrations of carbon dioxide on tree physiology and forest composition? How does elevated carbon dioxide affect soil organisms and soil nutrients? How is climatic change going to affect insect populations, fire frequency, and wind storm?

Coastal zones, in particular, are likely to reflect phenomena occurring over a larger geographical area. Already, the North Sea coast is showing the effects of the industrial and agricultural activity of central Europe, which is hundreds of mile away. The effects of industrialization may be catastrophic for many species, which may become extinct.

But a few species, and there is no way of predicting which ones, may thrive in the new environment. Therein lies the scientific importance of biodiversity: It increases the likelihood that at least some species will survive and give rise to new lineages that will replenish the Earth's biodiversity.

A global survey of diversity is essential to understanding the distribution of life forms on the planet. Biogeographical studies have shown that ecosystem diversity varies greatly from one region to another. Tropical forests, coral reefs, and deep benthic communities are very rich in species: deserts and high-latitude areas are species poor. Many studies have shown that species abundance changes naturally along latitudinal and humidity gradients.

An unprecedented transformation of the natural landscapes of the world is now occurring. Yet, different areas are being transformed at different rates. Some regions, such as the Mediterranean basin, have been altered by people for a long time. Other regions have been influenced only moderately by humans. Also, studies show that the response to disturbance varies greatly from one ecosystem to another. Therefore, it is important to determine how different biomes respond to human disturbance and to document the resulting change in biodiversity and ecosystem function.

Classical biogeography concentrated on the study of species ranges and the causes of their distribution. Given scientists' uncertainty about the number of species and the large number of undescribed species, new and more efficient approaches must be developed to complement past studies. These new approaches should take advantage of remote sensing technology and produce a geographic information system with statistical and analytical capabilities. However, because satellites cannot detect species diversity, new measures that are indicative of diversity and are more easily obtained by remote sensing must be used. For example, satellites can detect standing biomass and productivity, and research should help to establish the relationship between biodiversity and these measures.

Other problems to be studies include the effect that habitat fragmentation has on species ranges and on the probability of extinction and speciation; whether fragmented habitats are more prone to invasion; how invasions influence biodiversity; and the effect of increased geographical isolation on populations and species. Also, research should be conducted to develop a methodology for the systematic comparison of biodiversity across regions. To this end, a biogeographical information system for assessing diversity might use satellite information to identify assemblages of species at different geographical scales for long-term monitoring.

Until now, research in biodiversity has been undertaken by taxonomists at such institutions as the Smithsonian Institution, the Harvard University, Herbaria and Musea, the Missouri Botanical Garden, the Royal Botanical Gardens in Kew, the Musea d'Histoire Naturelle in Paris, and the Komarov Institute in Leningrad, which are located for the most part in the north temperate zone, while biodiversity is found mostly in tropical countries. Taxonomic institutions in both the north and the south have been in relative decline over the last 50 years. Staffs have not increased or have increased only slightly, and resources allocated to taxonomy have not grown in the same proportion as those devoted to other areas of biology. Consequently, at a time when the conceptual and practical importance of biodiversity is becoming clear, there is a shortage of trained personnel and a lack of necessary funds.

Taxonomists alone cannot answer the questions that have been raised by the massive transformation of the Earth. A collaborative effort by many kinds of biologists, including geneticists, physiologists, and ecologists, is required. As new techniques of genome analysis become routine, even molecular biologists may contribute to identifying genetic variation and its biological role.

Both governmental and non-governmental institutions should collaborate to advance scientific understanding of the significance of biodiversity, and many more resources will have to be made available to fund the collaborative research. The International Union of Biological Sciences, the Scientific Committee on Problems of the Environment, and the Man and the Biosphere Programme of the United Nations Educational, Scientific and Cultural Organization are spearheading research in biodiversity. These organizations are in close contact with the International Geosphere-Biosphere Programme and the United Nations Environmental Programme, to minimize overlap and to coordinate activities. Many national institutions also have active programmes of research in biodiversity, but most of these efforts suffer from inadequate funding.

Biodiversity research and conservation figure prominently in the agenda for the 1992 United Nations Conference on Environment and Development in Rio de Janeiro, Brazil, where conference planners are seeking a substantial increase in financial support for research in biodiversity.

It is imperative that biologists and the public learn more about the importance of biodiversity and its role in ecosystem functions. Only an international collaborative effort that is supported by adequate resources can accomplish this task. Increased public awareness of the serious implications of humanity's depletion of biodiversity before its importance is understood is helping to create a climate that may stimulate governments to support national and international efforts in this field.

Biodiversity is the variety of the world's living species, including their genetic diversity and the communities and ecosystems that they form. Today, this diversity of life is facing serious threats. Scientists concur that, at current rates of tropical forest loss and disturbance, roughly 5 to 10 per cent of all tropical forest species will be lost per decade during the next quarter century. The extinction crisis is not restricted to tropical forests. Freshwater habitats are being altered dramatically as rivers are impounded and exotic species introduced. In the southeastern United States, for instance, 40 to 50 per cent of the freshwater snails are extinct or near extinction.

Oceanic islands, where most of the extinctions of the past several thousand years have occurred, also remain among the most threatened ecosystems on Earth. Some 60 per cent of the plant species native to the Galapagos Islands are threatened with extinction.

Although the numbers of species that will be lost are greatest in tropical countries, the proportion of species at risk in the temperate zone is still large. In 1988, the Centre for Plant Conservation in Jamaica Plain, Massachusetts, estimated that more than 10 per cent of U.S. Plant species and subspecies are in danger of extinction in their wild habitats.

The loss of species and genetic diversity and the degradation of habitats and ecosystems represent an enormous cost to this generation and all future generations. Biological resources — the portion of biodiversity that is of actual or potential use to people— provide the basis for most human enterprises.

The species being lost today possess unknown food, medical, and industrial uses. The ecosystems being degraded through mismanagement of biological resources are losing their capacity to support the human populations dependent upon them, and this degradation is exacting further costs through soil erosion, salutation of reservoirs, local climate changes, desertification, and loss of productivity.

Also, the erosion of genetic diversity among agricultural crops through such practices as the substitution of high-response cultivators for local varieties hinders efforts to breed improved varieties. High response vultivars often handicap small farmers who cannot afford the necessary expensive inputs like fertilizer and pesticides.

The fundamental social, ethical, cultural, and economic values of biological diversity have long been recognized in religion, art, and literature. The great interest of children in nature, the numerous wildlife clubs, the generous donations made to conservation organizations, the political support for green parties, and the popularity of zoos and wildlife films are personal and economic expressions of preference that show that the public does not think of biological resources merely in terms of their cash value.

Nevertheless, policies that promote conservation can better complete for the attention of government and commercial decision makers if their advocates demonstrate the economic contribution that biological resources make to development. Even partial valuation of the benefits of conserving biological resources can provide, at least, a lower limit to the full range of benefits that conservation contributes to national economic development.

Three main approaches have been used to determine the value of biological resources:

- Consumptive-use valuation involves assessing the value of resources, such as firewood, fodder, and game meat, that are consumed directly, without passing through a market.
- Productive-use valuation involves assessing the value of products that are commercially harvested and marketed, such as timber, fish, game meat, ivory, and medicinal plants.
- Non consumptive-use valuation involves assessing the direct values of ecosystem functions, such as watershed protection, photosynthesis, climate regulation, and soil production, along with the intangible values of keeping options open for the future and the pleasure of knowing that certain species exist.

Such methods of valuation reveal that biological resources confer much larger economic benefits than is typically appreciated. For example, some 84 per cent of the Canadian population participates in wildlife-related recreational activities (a consumptive use), which provide Canadians with benefits that they declare to be worth $800 million (U.S.) annually.[6] In Sarawak, Malaysia, a detailed field study found that wild pigs harvested by hunters had an annual market value of about $100 million.

Policymakers often appreciate productive-use values more than consumptive - or non consumptive-use values relate to large industries, such as forestry and fishing, which are connected to markets and, therefore, reflected in national income accounts. Yet few people appreciate the magnitude of the contribution of biological resources to national economies.

One groundbreaking study that showed how to estimate the dollar value of biological resources and analyzed the contribution of wild plant and animal species to the U.S. economy concluded that some 4.5 per cent of the U.S. gross domestic product is attributable to wild species. In the late 1970s, 4.5 per cent of the U.S. gross domestic product was about $87 billion per year.

Although indirect values of biological resources are the most difficult to estimate, they actually may far outweigh direct values. Species without consumptive - or productive-use values may play important roles in the ecosystem and support species that are valued for their productive or consumptive uses.

For example, the retention of a wetlands complex outside Boston, Massachusetts, provides an annual savings of $17 million in flood protection alone — excluding the many other benefits, such as sediment reduction, fish and wildlife production, and aesthetic values, that the wetlands afford area residents. In Thailand, the $1.5-million cost of protecting the watershed of the Nam Pong Reservoir more than justified by the lessened rate of reservoir siltation, which requires expensive dredging.

Already enormous, the value of biological resources to humans will grow as those resources shrink through environmental degradation. Diversity in genes, species, and ecosystems provides the raw materials with which human communities adapt to change. Thus, the loss of each additional species, gene, and ecosystem reduces the ability of nature - and people - to adapt to the changing environment.Where does the blame for the loss of biodiversity and the degradation of biological resources lie? On the surface, the answer seems clear. The proximate causes include large-scale clearing and burning of forests, excessive harvesting of plants and animals, indiscriminate use of pesticides, draining and filling of wetlands, destructive fishing practice, air pollution, and the conservation of wildlands to agricultural and urban uses.

The root causes of biodiversity loss are found in basic economic, demographic, and political trends. These root causes include the human demand for commodities, such as tropical hardwoods, wildlife, fiber, and agricultural products; the growth in human population, which, even without proportional economic growth, places increasing demands on natural resources and ecosystem processes that are already impoverished and stressed; settlement policies that promote the migration of the growing unemployed labour forces to frontier zones; debt burdens that force governments to encourage the production of commodities that can earn foreign exchange; energy policies that encourage inefficiency in many nations and, in so doing, add to the burden of air pollutants and the risk of substantial global climate change; and inappropriate land tenure arrangements that discourage rural people from investing in substainable agricultural practices. When biodiversity loss is defined as a result of its immediate causes, the response is to take defensive and, often, confrontational actions, such as enacting laws, closing access to resources, and protecting additional areas. Such responses are necessary in times of rampant over-exploitation, but they are seldom really sufficient to change the underlying social and economic causes. Defining the problem in terms of its root causes can stimulate a more constructive response that seeks cooperative efforts to address the social and economic foundations of resources depletion. Six main obstacles make it hard to conserve biological diversity:

- National development programmes undervalue biological resources monetarily.

- Over exploitation of biological resources yields great profit for traders and manufacturers (who can externalize environmental costs) while impoverishing the local people who have few other sources of livelihood and who must pay the environmental costs of overexploitation.
- The species and ecosystems upon which human survival depends are still poorly known.
- Scientific research often does not meet the needs of resource and protected-area managers.
- Conservation activities tend to be focused too narrowly.
- Institutions assigned responsibility for conserving biodiversity have lacked sufficient financial and organizational resources to do the job.

These obstacles block progress in both developed and less developed countries. However, because the tropics harbour a major proportion of the planet's biological diversity and because overexploitation of resources in the tropics is linked to resource demand in the industrialized countries, the most pressing needs for action to conserve biodiversity are in developing countries. Developed countries depend on tropical resources for industrial materials, breeding material, pharmaceuticals, tourism destinations, and numerous other tangible and intangible benefits.

However, the exploitation of the tropics by the industrialized countries has yielded great benefits without commensurate investments in conservation and without compensation for the environmental costs of overexploitation. Such factors as cheap labour, artificially low-priced raw materials, inappropriate development aid, and the control of commodity prices and interest rates have encouraged rapid resource depletion and destruction. The situation is worsening continuously through the ramifications of the developing world's debt crisis and related high interest rates. Conservation will succeed only if both proximate and ultimate causes of environmental degradation and species loss are addressed. The complex threats to biological diversity call for a wide range of responses from many private and public sectors, and the mix of these responses must be adjusted to suit local conditions. Because government policies often are responsible for depleting biological resources, policy changes usually are a necessary first step toward conservation.

National policies that deal directly with wildlands management or forestry or that influence resource use indirectly by affecting, for example, land tenure, rural development, family planning, food prices, pesticide use, or fuel consumption can have significant impacts on the conservation of biodiversity. National and subnational conservation strategies can often provide the mechanism for identifying necessary changes.

The best way to protect species is to *protect habitats,* and most national governments have established legal means to do so. These means include creating national parks and other types of reserves (some 4,500 major-reserves exist ,covering nearly 500 million hectares); enacting local laws to protect particular forests, reefs, and wetlands; restricting private concession agreements; and establishing zoning regulations.

The responsibility for managing protected habitats often is spread widely among public and private institutions. Although accomplishments to date are impressive, the amount of habitat protected must be increased by a factor of three if it is to make the necessary contribution to the conservation of biological diversity.

Many of these new areas will need more flexible approaches to management than are customarily used in national parks. Protected habitats will succeed in realizing their conservation objectives only to the extent that the areas themselves are managed effectively and that the land surrounding them is managed in a way compatible with those objectives.

Implementing such management typically involves making protected areas parts of larger, regional schemes designed to conserve natural resources, induce sustainable human use of the resource base, ensure that benefits of the habitat protection reach the rural population, and involve the rural population in policy decisions related to the protected areas.

Ex situ preservation programmes, such as zoos, aquaria, seed banks, and botanical gardens, supplement in situ conservation by providing for the long-term storage, analysis, testing, and propagation of a few threatened and rare species of plants and animals and their propagules.

Ex situ preservation is particularly important for wild species whose populations are greatly reduced in numbers and serves as a backup to in situ conservation, as a source of material for reintroductions to the wild, and as a major repository of genetic material for future breeding programmes of domestic species. Some ex ritu facilities — notably zoos and botanical gardens — provide important opportunities for public education, and many make important contributions to taxonomy and field research.

Curbing the pollution of the biosphere is the biggest and most expensive conservation effort. Biological diversity is threatened by various forms of chemical pollutions, but ultimately, the gravest threat may be global climate change. Mean world temperatures could increase by about 2^0C in the next 40 years with an attendant rise in sea level of between 30 and 50 centimeters. Although the species and ecosystems within protected areas will certainly be affected by climate change, it is unrealistic to expect the boundaries of existing protected areas to be moved to track the changing climate because these areas usually are surrounded by more intensive human land uses. Instead, new forms of intervention will be required to maintain the ecosystems deemed desirable.

Many response to the loss of biological diversity have been supported by international legislation that has fostered global cooperation in conservation, as has the Convention on International Trade in Endangered Species, for example. However, species and ecosystems are still being exploited at rates that far exceed their sustainable yield. Recognizing the growing severity of threats to biological diversity and the increasingly international nature of the actions required to address the threats, the World Conservation Union(IUCN), based In gland, Switzerland, and the United Nations Environment Programme (UNEP), based in Nairobi, Kenya, have begun to prepare an international convention on the conservation of biological diversity. This effort has gained broad support from many governments. For example, in 1988, the U.S. Congress passed a joint resolution, H.J. Res. 648, in support of the conservation of biological diversity.

Any strategy to foster the sustainable use of biological resources must be based on public involvement. Indeed, initiatives that do not involve local communities generally are doomed to fail. Indigenous people in many parts of the world are especially reliant on natural resources for their cultural continuity and economic well-being. Their role in conservation should be given particular attention, and they should be given opportunities to participate as major players in the design of conservation programmes affecting their resources. Local people should be closely associated with the authorities responsible for both the management of biological resources and the establishment and management of protected areas.

Effective action must be based on accurate information, and the more widely shared the information, the more likely it is that individuals and institutions will agree on the definition of problems and solutions. Developing and using information is, therefore, an essential part of conservation at all scales, from the local to the global community. The current state of knowledge about species and ecosystems is woefully inadequate; detailed knowledge is still lacking on the distribution and population sizes of even such large and well-studied animals as African primates.

It is self-evident that increasing knowledge about the kind and variety of organisms that inhabit the Earth and the ways that these organisms relate to each other and to humans must be a foundation of conservation action. This basic inventory and research should be carried out simultaneously with field conservation efforts; the two activities can be mutually reinforcing. To gain the necessary knowledge, scientists must

- Conduct taxonomix surveys to document the wealth of the world's plant and animal species;
- Undertake greatly expanded ecological research to determine how the various pieces fit together, explain the population dynamics of species of particular concern, assess the effects of fragmentation of natural habitats, and determine what management steps are required to enable ecosystems to flourish with their full complements of species;

- develop and improve mechanisms for ex situ preservation, including both captive propagation of species and their eventual release into natural or restored ecosystems;
- monitor the changes in species diversity and ecosystem function as human influences become more pervasive;
- assess the ecological differences between relatively large but minimally disturbed ecosystems and ecosystems that have been heavily affected by humans to learn how to enhance productivity and restore degraded ecosystems to a more productive state; and
- conduct social science research to determine how local people manage their resources and how changes in resource availability and land use affect human behaviour.

Government agencies, local communities, and conservation organizations all need information to manage biological resources more effectively! Tools that can help meet this need include basic descriptions of flora and fauna, practical handbooks for field identification, rapid inventory techniques, and refined computer programmes for species and habitat inventory.

The need for information about the tropics is particularly important because these areas hold most of the world's biological diversity and are losing species faster than scientists can record them. The highest priority for basic inventory work should be given to the most threatened sites of greatest diversity and local endemism. The information contained in the species in these areas could disappear before humanity even knows what it is losing.

Development agencies should support national efforts to establish local, sectorial, and national information management systems by demonstrating methodologies, providing training opportunities for taxonomists and biologists, and subsidizing the publication of status reports. Universities, research institutions, and nongovernmental organizations need to be strengthened so that they can help governments assess their biological resources. Closer working relationships should be established between museums and other taxonomic-oriented institutions and those concerned with conservation of biological diversity.

Is there anything in common between the familiar *neem tree* and a heated global debate over protecting the environment? Or between its overwhelming medicinal and pesticidal attributes and a multi-million dollar industry in a rich and developed nation? Yes, surprisingly. For this hardy tree, taken for granted as part of the Indian landscape, is the focus of scientific research in the USA because of its amazing array of *pesticidal properties*. And as studies exploit the goodness of neem via biotechnology, it is likely that somewhere along the line the indegenous people who discovered the bitter magic of this tree will be forgotten, just as its genes begin to show their might in the dollar avatar. Is this an instance of the sharing of a global resource?

The debate over this biological wealth deepens because most of it, on a global scale, originates in the developing, tropical nations of the Third World, which seem to be at cross-purposes with the developed nations who have adopted a big-brother attitude. The rather verbose draft convention has the laudable objectives of conserving biological diversity by ensuring sustainable resource use, providing funds and transfer of technology-to developing nations for conservation, and the fair sharing of the benefits of research in biotechnology arising from conservation of biodiversity. Lengthy and longwinded, the draft has 40 articles, while the overuse of technical jargon coupled with convoluted sentences makes it almost incomprehensible. In the absence of a complete understanding of all its implications, India, as one of the world's biologically wealthy nations, is treading softly, even as the minister for environment and forests, Mr. Kamal Nath, is voicing his objections. But what does India stand to gain from the Convention? Will it protect India's biodiversity from the new global order of bio-technological imperialism? India certainly has a lot to protect. "Our country is extremely rich in genetic resources with about 15,000 known species of flowering plants," says Mr.R.S. Rao, Director, National Bureau for Plant Genetic Resources, New Delhi, the country's premier institute devoted to gene conservation and collection.

Imagine, then, the amazing array of candidate genes that must be available for biotechnological manipulation. Says Swarahj who heads the insect research division at the Indian Agricultural Research Institute, New Delhi, "Insects constitute a mind-boggling 80 per cent chunk of all living organisms in India, there being roughly 60,000 different species.

Many of these, like the moth which weaves silk, have revolutionized human civilisation. Then there are the microbes, bacteria, fungi and the multitude of sea life, all vital sources of medicines and life-giving compounds." The list is endless, chock-full of incredible genes from nature, which is why India wishes to understand fully what she stands to achieve from the Biodiversity Convention.

Technology transfers and financial assistance of course, say the pundits. It is this link between biotechnology and bioresources that has reached some consensus now, says Mr. Kamal Nath in an exclusive interview soon after his return from the update meeting in New York. It's here that the trouble begins as the patronizing attitude of the West could well mean a 'recolonisation of India'.

Points out Ashish Kothari, an observer from the non-governmental sector at the Convention's last meeting in Nairobi, "Technology should be exchanged and not transferred one way since there are a number of appropriate, indigenous technologies which are already underway in India." Of course, there's no doubt that India's level of technological development is strong. Even so Mr.V.L. Chopra, Director General, Indian Council for Agricultural Research, New Delhi, says, "It may be wrong to take a blanket view and reject all foreign assistance with regard to technology.

Instead there should be a case-by-case assessment of the technologies being transferred with a view to retaining the sound and rejecting the useless. Says Mr. Rana "There are two categories of nations — the 'gene-rich, technology-poor' and the 'gene-poor, technology - rich' nations. India falls into neither of these categories being 'gene-rich and technology-rich.

Why then the complications of this convention? "The very basis of biological wealth," says Mr.Chopra, "has been the general concept that the free flow of germplasm on an international scale should continue." But who will call the shots? Chhatrapati Singh of the Indian Law Institute, New Delhi, an expert on environment law, feels that the convention is yet another example of Western formulation of international agreements which, far from being globally participatory, are mere 'sign-on-the-dotted-line' documents. Also, the sharing of biological wealth must be judged in totality along with the issues of intellectual property rights, the GATT and patents. Even Mr Kamal Nath believes in the need to protect our biological wealth via IPRs and patents. Here one comes full circle to the neem tree. India is a treasure-house of living things and has a wealth of genetic resources in its umpteen biographical zones — the deserts, coastal areas, rain forests, teak and sal forests and the meadowlands of the mountains. Does their conservation really require a Convention? Yes, say spokespersons of the ministry of environment and forests, but not a convention which will cost the nation its sovereignty. There is also the feeling that the Convention will focus more on the utilisation of resources than their actual conservation.

Can we then set our own terms without foregoing our independence? Says Mr.Rana, "One way to check the imperialism being forced on biologically wealthy nations like India is to strengthen our gene banks and research base, inventorise and characterize our genetic wealth and invest it in basic research."

Realising that there are several shortfalls in government programmes, M.S. Swaminathan, President, World Wide Fund for Nature (WWF), launched a people's biodiversity programme. Under the aegis of the Jawaharlal Nehru Community Biodiversity Conservation Movement at the WWF, these projects have managed to involve effectively the local communities in reforestation, nursery work and wildlife conservation programmes all over India. Several such programmes have been highly successful. Some even suspect the motivation of the unusually generous offers made by certain developed nations with regard to financial technological assistance.

Even as Mr.Kamal Nath declares India's decision to reject the Convention unless suggested changes are made, India might continue to attend the negotiations and pre-

Rio meetings. Will India be lured by these attractive offers of monetary help? Or will she continue to guard her national interests and biodiversity from the developed nations' hegemony? Though Mr Kamal Nath protests that India is not on dole, only time will tell how these questions will be answered ultimately. Meanwhile, the draft convention high-handedly proclaims awards to the originators of traditional methods or using biological wealth. In reality, there are so many illegal outlets for both indigenous genes and their utilisaticn methods that losses occur every day.

So, if neem is today the catalyst of a dollar machine, it may have been another plant yesterday and might be yet another tomorrow. A faulty conservation strategy with one-sided formulations may only further the damage, making poorer nations the 'beneficiaries' of low-grade, obsolete technologies. These losses may far outweigh the gains from such a Convention.

Yet there are the silver linings. Mr. Kothari feels that with negotiation and substantial modification, the Convention could be a useful tool for the conservation of biodiversity since it would provide a vital link between nations, with proposed global lists avoiding duplication while ensuring the conservation of all species. " The balanced discussions between the 80 participating countries have been a definite step forward in the chequered history of environmental movements," he points out. "Further, we should adopt a holistic view of the Convention without rejecting it outright. After all, one never knows when India might be in dire need of a gene not to be found here." Though the facts are clear, the implications are not. There are too many spokes in the wheel. One worrying aspect is that a great majority of people — even people in key positions — seems to be ignorant of the convention's text.

Its political understones are strong and could prove overwhelming. The fault is perhaps in the preliminary formulation of the draft carried out at the World Resources Institute, Washington, DC. If participatory drafting involving all members — 'contracting parties' as the Convention calls them — had been done, some of the hurdles could have been avoided. And yet, Mr. Kamal Nath is of the opinion that with the G-77 group of developing countries backing India's stand, there is room for optimism. Meanwhile, the 'fair and equitable sharing of resources' seems to have become the golden phrase now. So Utopian does this seem that one would imagine it will never materialise. Earlier instances of gene hunting in the tropics by Western nations are far from being rosy stories of North-South cooperation. Perhaps, the goodness of neem needs protection in more ways than one.

Although there are a number of agroecosystem management approaches which provide alternatives to these trends, Bio-Dynamic Agriculture, Fukuoka's Natural Farming , organic agriculture 'alternative' agriculture sustainable' agriculture 'ecological' agriculture Holistic Resource Management is of particular interest in the context of biodiversity issues because of its emphasis on biodiversity protection and enhancement as a central management principle.

In Holistic Resource Management courses, students are encouraged to work with fundamental ecosystem processes and increase biodiversity on their farms or ranches to enhance profitable sustainability. Savory builds his argument for greater biodiversity in agroecosystems on the debated hypothesis in ecology that stability is positively correlated with diversity in ecosystem function. McNauhton's work in old fields, East African grasslands, and more recent work by Tilman and Moffat support Savory's contention.

The fundamental ecological premise of HRM is that efficient nutrient cycling, hydrology, successional dynamics, and energy flow, with a strong biodiversity base both above and below ground, buffers managed ecosystems' responses to perturbations. Theoretically, by learning to work with and improve biodiversity and these ecological processes on their land, farmers and ranchers need less external inputs to sustain production, which then allows them to reduce their input costs, enhance their profitability and sustain their resource base for the long-term future.

It is our idea that biodiversity, used as a management concept, could impact ecological processes at population, community and ecosystem levels. For example, while diversified cropping practices, in both space and time, can contribute to increased populations of natural enemies, the same practices offer opportunities to recycle nutrients within a farming system and, also, present opportunities to value add enterprises within a farm.

Perhaps most fundamentally, a decision process such as HRM provides a framework that encompasses a diversity of practices, a ranee of options that a farmer can then 1 best adapt to local environmental, economic and social conditions. For these reasons, we believe it is important to view biodiversity as a multi-function management tool. We desired to learn how exposure to HRM influences attitudes toward biodiversity among farmers and whether the actual experiences of HRM practitioners support Savory's idea that biodiversity is the foundation of profitable sustainability in agriculture. We used a participatory ethnographic approach and qualitative research methods to explore the attitudes, perceptions, decision making, observations and experiences of a group of farmers and ranchers who had proven expertise in using HRM.

Our data are based on the practical experience of these land mangers rather than experimental scientific data. Cheekland and Bawden et al. describe two different approaches to agricultural research. The *scientia* approach is quantitative, focused, precise, slow, expensive and, if done in optimal conditions, so controlled as to be disassociated from the complexities of reality. It focuses on mechanisms and predictability. In contrast, the praxis approach is qualitative, rich in contextual information, relatively inexpensive, quick and highly confounded with complexities of the real world.

While many agricultural scientists are accustomed to scientia information, a growing number are recognizing the value of the kind of information produced in praxis studies, which can provide critical insights into the context of a particular area of scientific inquiry.Although biodiversity may be viewed as a fundamental principle of sustainablity by many scientists and environmental activists, in our experience with farmers in Ohio, Missouri, Kansas. West Virginia and Vermont, at least, it is rarely taken into consideration. If we wish to develop more forms of agriculture which protect biodiversity, we need to understand the perspectives of the people who manage agricultural lands in general.

However, in particular, understanding the perceptions and experiences of land managers who are committed to these principles can be a valuable step towards developing research and education for a wider audience. Other researchers working in a participatory mode with farmers throughout the world have found that indigenous experience and knowledge have provided an effective foundation for research and development of more sustainable agricultural practices.

The Holistic Resource Management process is being used by farmers, ranchers, communities, government agencies and, increasingly, businesses, nationally and internationally. Holistic Resource Management evolved out of Savory's concern for the alarming rate of desertification he observed as a wildlife biologist in his native Zimbabwe and many other parts of the world. Savory equates desertification with loss of biodiversity. HRM students are taught that biodiversity is the source of prosperity and that loss of biodiversity can have far-ranging environmental and social effects. 'Biodiversity is not about rare and endangered species, it is about human survival'. Savory argues that the fundamental cause of biodiversity loss is the process of how management decisions are made and he offers the Holistic Resource Management Model as an alternative decision making process.

The first step in the HRM process is to define thee 1 whole' being managed in terms of people, landbase and money. Once a clear assessment has been made in these terms, the people directly involved should work together to develop a holistic goal. This goal includes: (1) their quality of life values, (2) forms of production they must achieve from the land or other resource base to support each of their quality of life values, and (3) a vision of what they wish the land or resource base to look like in the future to sustain their production and that of future generations.

In the future landscape description, HRM students are taught that it is important to understand and work with four ecological processes as foundation blocks community dynamics, the water cycle, mineral or nutrient cycles, and energy flow. Only after this contextual groundwork has been laid do specific tools come into the HRM process.

Although most people, and especially natural scientists, are much more comfortable discussing tools than quality of life values, it is tools that we most often argue about. Laying the context of a farm or ranch first can be a key to building, unity in otherwise disparate groups of people in the whole. The tools used in the HRM model include: human creativity, money and labour, rest, fire, razing, animal impact, living organisms, and technology. Creative brainstorming or idea generation is encouraged to produce a list of possible solutions to management challenges. These, then, are tested using seven testing guidelines in a pass or fail mode, against the established goal.

— The first testing guideline, Whole Ecosystem, asks how the proposed tool or enterprise will affect the four ecosystem processes; will they be enhanced or degraded and will it move these processes towards or away from the future landscape description?

— In the second testing guideline, Weak Link, the user determines the weakest link in their operation. Is there a lack of knowledge or information that is holding back progress toward the goal? Is the weakest link in personnel or financial problems, inadequate or poor land? Where is the weakest link in the energy conversion chain from sun to plants, to products that can be marketed? Is there a problem organism involved? If so, where is the weakest link in the life cycle of the organism?

— In the third testing guideline, Cause and Effect, the user considers whether the proposed action will treat a cause or symptom.

— The fourth testing guideline, Marginal Reaction, is one of the economic guidelines and is used only when comparing more than one option. It asks which option will provide the biggest return for the money or time invested in moving the operation towards the goal.

— The fifth testing guideline, Energy/Wealth Source and Use asks: Will the proposed tool require the use of finite sources of energy and will such use have to be repeated? The fifth testing guideline asks also: What is the source of money required paper dollars, mineral dollars (money from non renewable resources) or solar dollars.

— The sixth guideline, Society and Culture, asks how the proposed tool or action will affect thc culture and society. Will it help strengthen the community?

— The seventh and final testing guideline, Gross Margin Analysis, is another economic test and, like Marginal Reaction, compares several options., Using the approach of British economists, Wallace and Burr, in which only variable costs are figured, this guideline helps determine which forms of production are most profitable.

If used conscientiously, this decision making process gives the decision maker a consistent and objective evaluation of the soundness of the decision and helps insure that any decision is optimized between environmental, economic and social considerations. Savory argues that it is this conscious optimization process that sets HRM decision making apart from decision making that humans have been using throughout our history as a species and regardless of culture. The Management Guidelines part of the model deals With specific and more advanced aspects of Holistic Resource Management. See Savory and Bingham and Savory for additional information. In the last part of the model Control, Test Assume Wrong, Early Warning Criteria, Monitor, etc., the user assumes the decision is wrong and watches for the earliest sign of deviation from the plan, be this economically, ecologically or with respect to quality of life. This, approach creates a different psychology from that in conventional decision making in which a carefully researched decision generally is assumed to be right. The names and contacts of committed Holistic Research Management practitioners were requested from established certified HRM educators and the Center for Holistic Resource Management in Albuquerque, New Mexico. We wanted to interview a range of HRM practitioners with respect to several parameters: (i) length of time using Holistic Management, (ii) different climatic regions or position on the 'brittleness scale', (iii) scale of operation, and (iv) enterprises.

A set of questions was agreed upon. Potential interviewees were contacted by correspondence and given the opportunity to think about the questions before being called and interviewed. A few of the interviewees chose to respond in written form and four individuals were interviewed in person, but the majority were interviewed by telephone. The interviews took an average of one hour to conduct. Biological monitoring is the part of Holistic Resource Management which detects early warning signs of ecological problems. Savory and Bingharn and Savory present methods in which attributes concerning plant cover, density and species composition and soil surface conditions are measured on fixed transects. Instead of direct statistical analysis, the data are related to the HRM model for interpretation.

In addition to providing information on changes in plant community composition over time, this monitoring offers the manager a relatively simple and inexpensive means (in comparison with scientia research methods) of assessing feedback on parameters associated with all the ecosystem foundation blocks in the model.

Random sampling points are determined along fixed starting point transects using a dart thrown over the shoulder. Sampling is to be done at the same time of the year each year by the same person. Less than 5% of the people interviewed were doing the full scale biological monitoring recommended by Savory and co-workers. Most of them relied on less time-consuming methods of monitoring their land ecologically, which included walking their land and getting down to ground level and observing soil and plants.

Observations of faster decomposition of manure piles, more earthworms and other soil macro-fauna and changes in organic matter or soil tilth, for example, were viewed as indications of changes in nutrient cycles. Many interviewees reported observing changes in soil erosion and hydrological cycles, such as springs and streams running where none bad run for many years and clear instead of murky water in ponds. Another indication used for changes in the water cycle was both resistance and resilience to drought.

Finally, for changes in energy flow, we used reported increases in plant biomass and increase in stocking rates of livestock. Stocking rates of livestock on a unit of land was a parameter reported by numerous interviewees and was viewed as a direct indicator of profitability in addition to an indirect indicator of energy flow. Stocking rates are the amount of land required to support one animal for a specified period of time, usually a year.

The scale of operations ranged from 7.3 ha to 90000 ha and the time practising HRM 1.517 years with an average of 5.6 years. The cases were characterized by wide range of farm/ranch types. All but one of the cases were using planned intensive grazing on some of their land, if not all. Interviewees typically defined biodiversity as all the different species of plants and animals, including microorganisms in the soil, and their quantities as biomass or numbers. In addition, habitat diversity and genetic diversity within agricultural species were often mentioned.

One interviewee included diversity of human ideas in his definition of biodiversity. Nine percent of the interviewees thought about biodiversity in the context of their operations before being exposed to HRM. Now, 100% of them think biodiversity is important to the ecological and economic wellbeing, of their farms or ranches. 'Biodiversity is the true wealth to sustain the world. It is the only true source of capital for enterprises.

We were not aware of biodiversity before HRM, said one of the interviewees in response to question 6. All but one interviewee reported observing increases in biodiversity since they began using HRM. With respect to changes in profit, 80% reported increases from their land since HRM began influencing their decisions. Of these, 40% provided actual percentage increases which ranged from 60% to 1400% and averaged 386%. Sixteen percent reported no or little increase in profits yet, but were optimistic about future increased profitability.

In addition to reported increased biodiversity on their farms and ranches and concomitant reported increases in profitability, almost all of the interviewees reported observing improvements in ecosystem processes since they bean using HRM, For example, a farmer/ rancher in North Dakota with 10 years' experience practising HRM, related increases in soil permeability and infiltration from 5 em to 50 em. All of the ranchers west of the Mississippi River reported changes in plant species composition to a greater frequency of perennials and the return of many native tall and short grass prairie species.

On the human community side, most of the interviewees relayed that they were considered odd by their immediate neighbours, but many of them are sought after speakers for national meetings and even internationally. Networks with other HRM practitioners were considered critical to the success of most of the interviewees. With respect to quality of life issues, 91% of the interviewees reported that HRM has influenced some aspect of time management in their operations and lives in general, of this group, 29% reported no change in total labour, but that how that time is allocated has changed, 'I don't know that the amount of time we spend has changed, but how we spend it has', said one of the interviewees. 'We are no longer doing a lot of things we do not like to do', reported another. Fifty-two percent, reported decreases of up to 40%-60% in labour requirements in their operations, in spite of the extra planning and monitoring required by HRM. 'Not only do I have time to go out to eat with my family one night a week now, but I can pick which night', one of the ranchers told a group of HRM workshop participants. Another rancher reported, 'I have time to be a dilettante'. All of the ranchers interviewed on the original tall and short grass prairies of the USA reported the return of native species in their intensively managed grazing cells and reported that their cattle were doing well on the native forage.

Ecologically, this is very interesting, in that it was agriculture that caused degradation of the American prairies and on these HRM managed lands at least, agriculture is helping to restore the native ecological communities. These ranchers believe that their investments of time and money to upgrade their land is netting increased profits in higher carrying capacity and lower production costs. These three operations offer a range along Savory's brittleness scale, with the farm in West Virginia in the humid southeastern part of the USA being in the least brittle environment, followed by the

Oklahoma Coffey Ranch, and the New Mexico ranch in the and southwestern part of the USA in the most brittle environment. The Fichtner family moved to this 32 ha hillside farm in 1981. At that time, Fichtner reported that the farm was almost overrun with multiflora rose and its soils were classified as severely eroded and very severely eroded. It had been plowed for maize production in the 1930s and then put into sod in the 1940s or 1950s and the land supported a few horses and beef cattle since the 1950s. The farm is located in an area which was fire-maintained savanna grazed by the Eastern Woodland Bison before European settlement. The Fichtners began intensive grazing in 1989 and discovered HIM at an intensive grazing conference The husband and wife took their first HRM classes in 1990 and 1991, respectively. Both had off-farm jobs at that time. See Table 3 for their goal statement. With their goal and the principles and tools of HRM they began to reclaim their land.

A great diversity of animals are used as tools each with their own role. For example, the hogs break up and compost manure in the barn and the Muscovy ducks control flies. The Scottish Highlander cattle are very rugged and excellent browsers clearing brush efficiently. The donkeys keep coyote at bay. The cattle graze after the sheep, helping to break the parasitic cycle, as well a harvest the plants their predecessors passed over Lambs are pasture born in the spring. The Fichtner began controlled grazing in 1990 on a small seal with seven paddocks, a few sheep and electric net on a 0.2 ha field. Now they have one 12 ha cell with 1 permanent paddocks, which they further subdivide into 40 or 50 or even more temporary paddock depending on the conditions. In this time, profits from the farm rose enough that the wife was able to quit her off-farm job.

2 Biodiversity of Social Biota

In this chapter we consider studies concerned with soil fauna biodiversity in chernozem soil in a set of ecosystems ranging from natural, undisturbed ecosystems, to intensively manipulated agroecosystems to investigate the mechanisms of biodiversity maintenance and methods for their estimation. Most of the results of the studies covered here were published, in detail elsewhere.

Soil macrofauna biodiversity was studied during the May-October periods of the years 1978-1991 in the following set of ecosystems located in the territory of the Central Chernozem Reserve and the Kursk Agriculture Station: 120 ha of natural unmown meadow steppe; 500 ha of mown meadow steppe; 60 ha of pasture 45 ha of barley field without fertilisation 1 ha of permanent black fallow ploughed twice every year

The mowing cycle in the mown steppe is hay-making once per year for three consecutive years followed by one year without hay-making. The meadow steppe and pasture ecosystems are the only grassland ecosystems in the Central Russian region that have never been ploughed during human history. They were used as horse pasture or hay-making land until the 1930s, when they became part of a reserve. Other small islands of real steppe can be found in Ukraine. These ecosystems were thoroughly investigated by climatologists, soil scientists, botanists, geochemists, vertebrate zoologists and others for long periods of time, and a lot of data have been published on the chemical composition and long-term dynamics of the ecosystems. Plant biodiversity studies have been carried out for more than 60 years while soil animal biodiversity research began about 40 years ago. The main studies on soil chemistry were carried out in 1950 1980, plant chemistry in 1950-1980 and animal chemistry in 1970-1980 in the same ecosystems. In most of our studies we used the following method of soil macrofauna estimation.

Soil samples were taken by soil corer up to 30 cm depth. Every 10-cm layer core was put into a plastic bag, brought to laboratory and fauna was extracted by hand sorting. After the extraction, animals were kept on a moist filter paper for 24 h to excrete the gut content. They were then dried at 100^{0}C in a thermobox for 48 h and weighed. Methods of chemical analysis of the soil, plants and animals collected have been described in detail elsewhere. Earlier we had collected soil samples of 50 cm X 50 cm X 30 cm and of 25 cm X 25 cm X 30 cm and extracted animals by hand sorting in the field. However, it was found that collection of these large samples led to disturbance of the vegetation cover around every sample of an area four to six times larger than the area of the sample.

This sampling procedure could change the vegetation cover in and around places of collection and alter natural processes in the unique reserve of ecosystems (we collected 10 samples every 2 weeks from May to October in the unmown steppe and a reserve oak forest). It was therefore decided to modify the sampling procedure, i.e. sample size and number of samples. Twenty samples of 10 cm X 10 cm every 2 weeks were considered sufficient for the study of seasonal dynamics of soil macrofauna in the reserve habitats. Population numbers of different species of soil macrofauna obtained by this sampling procedure were practically equal to those found with the earlier method. The total number of soil samples collected during the period of investigation was near 3000. This includes all samples gathered during studies cited below. The species number and population characters of soil macrofauna decreased in the following order: unwown steppe > mown steppe > pasture > barley field > permanent black fallow. It is noteworthy that in the black fallow ecosystem only one or two species from any of the macrofauna groups were found, namely the earthworm Nicodrilus caligonosus, the diplopods Brachyiulus jawlowskii and Megaphyllum rossicum, the staphylinid beetle Amischa analis, and the carabid beetles Bembidion lampros and Microlestes minutulus.

None of the geophilids or lithobiids that are common in the other ecosystems were found in the black fallow ecosystem. The elaterid larvae found in the soils studied were phytovorous or carnivorous, and it is therefore evident that they were absent in the black fallow soil. Only ubiquist species were present in the barley field (the same species of earthworms, diplopods and carabids as in the black fallow soil, as well as the staphilinid beetles *Scopaeus minutus* and *Amischa analis,* the elaterid larvae *Agriotes sputator,* and the lithobiid *Monotarsobius curtipes),* and super-dominance is the particular feature of the populations in the barley field and black fallow soils.

The exception are geophilids species: *Geophylus proximus* comprises 67% of all geophilid specimens, and *Arcto-geophilus macrocephalus* and *Schendula nomorensis* 16% each. Practically all animals found in the field and the fallow soil are microbivorous or carnivorous. Although grazing pressure at the pasture is relatively low, and along the pasture borders there are patches of steppe vegetation, soil macrofauna biodi-versity is low in the pasture compared with the steppe ecosystems., Only the carabids and elaterids show an increased species number in the pasture soil compared with the barley field and black fallow.

Amara species of carabids and wireworms are new members of the groups under study in the pasture soil, in addition to the species already mentioned for the barley field and black fallow. Staphylinid beetles found in the pasture soil were all coprofillous.

Regarding geophilids, *Pachyinerium ferrugineum* replaces G. *proximus* and S. *nemorensis* found in the barley field soil. Density and biomass of the soil macrofauna population in the pasture soil is similar to that in the soil of mown steppe, but biodiversity is lower. Biodiversity is notably greater in the steppe ecosystems compared with the other ecosystems studied and is comparable in the mown and unmown steppe plots.

However, density and biomass are markedly higher in the unknown steppe than in the mown steppe. There are no differences in the species composition of earthworms, diplopods, geophilids, lithobiids and elaterid larvae identified in soils of the mown and unmown steppe.

Some new species appear in these groups in addition to those mentioned above. The differences are marked only for population density and biomass of separate species in these groups. E. *nordenskioldi* is the dominant species in the unmown steppe soil and N. *roseus* in the mown steppe soil. On the other hand, the staphylinids and carabids differ markedly in these ecosystems. The Soerensen index for carabids is 0.64 and for staphylinids is 0.52. The Jaccard index is 0.47 and 0.36, respectively.

All other community parameters, such as mean species number per sample, mean density and biomass of animal populations, are correlated with the changes of biodiversity (Table 2). Mean species number per sample shows a stronger correlation with the population density and biomass than does total species number. This is because mean species number per sample is a characteristic of dominant and common species in determined density and biomass. A decrease of biodiversity of soil mesofauna groups is a common phenomenon in sets of ecosystems with increasing human influence.

Among the many possible reasons, soil perturbation, and microclimatic and soil chemistry changes are considered the leading causes Even a low-impact human activity such as hay-making is reflected in the soil macrofauna composition, and differences between the mown and unmown steppe communities are considerable.

Pitfall data collected in the same period indicate that in the mown steppe the share of ubiquists among staphilinid beetles and carabid beetles is 17% and 5%, respectively, in the unmown steppe it is 28% and 7%, respectively, in the pasture 32% and 44%, respectively, and in the barley field and black fallow ecosystems 100% for both groups. It is interesting that the share of ubiquists in the field increased from 72% in the 1960s to 100%. This may be a result of an increase in the human impact on the local environment as a whole.

In our opinion the higher proportion of ubiquist species of beetles in the unmown steppe is related to the microclimatic conditions in the thick layer of, litter, where conditions are optimal for ubiquist species. On the other hand, *E. nordenskioldi is* the only non-ubiquist species among earthworms in both ecosystems but is dominant in the unmown steppe. Two other species are found in all anthropogenic habitats and may be considered as ubiquists, with *N. roseus* dominating in the mown steppe. Based on data on biodiversity, it is difficult to estimate what type of a reserve regime is closer to the initial prehistoric conditions of real steppe (mown or unmown steppe).

The grass litter layer is very thick in the =mown steppe but practically absent in the mown steppe. Geobotanists consider pasture regimes to be the initial one for real steppe, and the mown steppe regime as analogous to the pasture (grazing) regime. However it is evident then that pasture and mown steppe cannot be considered as analogues.

The biodiversity and community characters prove this assumption. Grazing activity in pasture at the level of one cow per hectare is sufficiently high to disturb steppe ecosystems and change them entirely. It is very significant that this level of pasture activity does lead to the disappearance of the grass litter layer. Many pasture soil species are ubiquists or coprofillous.

Our observations showed that such coprophagous beetles as Aphodius and Onthopha*gus* species are common in surface soil of pasture. These species are practically absent in the mown and unmown steppe

The point of view that pastures and mown steppe are analogous ecosystems was universally adopted by geobotanists. However, a study of species composition at a steppe plot after artificial fire shows that during the first 2 years soil fauna species composition is similar to that on mown steppe plots. Then species composition is changing toward that typical of unmown steppe. It confirms our opinion that steppe ecosystems are cyclicpyrogenic ecosystems, and not pasture ecosystems.

Therefore the species composition of soil macrofauna in the mown and unmown steppe denotes that these ecosystems can be regarded as separate stadia of post-fire succession series in meadow steppe, and not as an initial state of real steppe ecosystems. Earlier we showed that changes in biomass and production of soil animal populations in agroecosystems depend on available organic matter and nutrients in arable soils Correlation between species composition and population density and biomass suggests that resources of nutrients also have an influence on biodiversity. The differences in composition and biomass of the soil fauna in arable plots are small although the quantity of organic resources in the field soil are somewhat higher than in the fallow soil In the fallow soil there are no such sources of available carbon, nitrogen and phosphorus as plant roots for animals and microorganisms. Available organic matter for animals in field and fallow soils is mainly present in the form of decomposable humic compounds and microorganisms. The low biodiversity in field and fallow soils are probably a reflection of small pools of available organic matter, nitrogen and phosphorus sources for the fauna. Stores of available organic matter in the natural steppe are three times higher than in the field and five times higher than in the fallow. The microgen pool in the steppe is 1.5 times higher than in the field and fallow. The phosphorus pools in the different ecosystems are practically equal. However microbial biomass in the soil and in the ~tits of soil animals are the main sources of essential nutrients for animals Biornass of' microorganisms in field and fallow soils is higher than in meadow steppe soil, but it is only 'potential' biomass. To be the nutrient pool for animals, in microorganisms need stores of available energy in soils. These stores are concentrated in dead plant residues and plant exudates. Soil microorganisms in the field and fallow are active if there is an input of available carbon compounds.

Cellulolytic activity of microorganisms in soil of the fallow was higher than the activity of both soil fauna and microorganisms in soil of the meadow steppe when dead roots or hay were put into the soil. Cellulose for fallow microorganisms was a good source of available energy, while steppe microorganisms probably used more easily available sources of energy. Hence the biodiversity of soil fauna is possibly correlated with amounts of available organic matter and nutrients, and not only with habitat conditions. In arable soil it will be dependent on input of available carbon compounds for soil microorganisms. If we compare overall biodiversity of ecosystems we usually only consider biodiversity of multicellular organisms and not the biodiversity of microorganisms.

Unfortunately, there are some problems with the comparison of biodiversity of taxa of different size levels, and of ecosystems as a whole. Over 10 years ago, we first suggested a new index for biodiversity comparisons. Our aim was to find indicators for a more objective estimation of biodiversity. In our opinion, DNA in ecosystems could be an indicator of 'potential' biodiversity in an ecosystem as a whole. We called the suggested character the 'index of genetic saturation' (ICS) IGS is equal to the sum S(mjx kj), where S is the sum of the products of mj and *kj* for species of a separate group (S is equal to the species number in the group) in an ecosystem, mi is the production of the species of a group (kg ha and *kj* is the DNA concentration in this jth species. A preliminary comparison of IGS on the basis of data collected in our and other studies and known data on production and DNA concentration shows that agroecosystems are not very different from natural systems fir terms of IGS. The data have a speculative and preliminary character, and more precise and detailed studies are needed.

3 Living Resources

It is not just the wild living resources that are today threatened because of the destruction of their habitat, domestic species of plants and animals are also facing genetic attrition, partly because of those very activities that are undertaken by human beings in the name of economic development, such as the Green Revolution. Indian farmers probably grew over 30,000 varieties of rice. But with high yielding varieties becoming popular, these varieties are disappearing, and by the end of the century there will be no more than 50 varieties being used, with the top ten alone occupying over 75 percent of the acreage. This genetic erosion, if carried too far, could thereaten the very future of the human society.

Protection and conservation of living species including domestic species, must, therefore, become an important issue for public policy and public pressure. In recent years, the government has begun to take conservation of wildlife seriously. But the manner in which these conservation programmes are being conceived and implemented is creating more tension than peace and harmony. Whereas conservation is on issue of crucial importance, it cannot be undertaken without the involvement of the people, especially those directly affected. Nature conservation programmes cannot be promoted in the same ham-handed way as those programmes which have destroyed nature.

Plant and animal life, as it has evolved on this earth, has produced a bewildering diversity. There are millions of species of plants and animals, but even within a species, there exists enormous diversity. For instance, there are an estimated 1,20,000 types of rice plants alone in the world and this is only a guess. There can actually be many more. This massive diversity, or "gene pool", as scientists like to call it, is a common heritage of all human beings. It is also their insurance for future survival. Increasingly, as new varieties of plants and animals useful to human beings are bread, this massive gene bank provided by nature becomes important. Breeding is essentially an exercise in finding the desired gene from nature's vast genetic reservior and introducing into the genes normally used. The Green Revolution, for instance is essentially based on a strategy of genetic reconstruction aimed at producing new varieties of cereal plants which can produce high yields. Scientists involved in this work, however, have also realised that the Green Revolution is something akin to riding a tiger; once you are on it, you have to stay on it to survive. To keep the Green Revolution going, breeders must produce new varieties every five to seven years as older high-yielding varieties succumb to pests and diseases, which themselves keep changing in intesity and character. If one variety covers too large an area for too long a time, it becomes an open invitation to disaster. A disease can sweep through a genetically uniform crop like wildfire leading to widespread crop failure and famine. The brown rust infection, for instance, affected large areas in north India planted under the wheat variety kalyansona during 1973-74. Other than high yeilds, a high priority for both plant and animal breeders is the development of resistance to diseases, and screening available plant and animal varieties for these traits accounts for substantial amount of time and resources of breeders.

Ironically using "superior" genes on a massive scale to increase production, as exemplified by the Green Revolution in food crops, blue revolution in fish production, white revolution in milk, and massive afforestation with high yielding monocultures, has posed one of the biggest threats to the world's genetic diversity in livestock, fish, crops and trees. In each of these areas, human beings are promoting a narrow genetic base-a few genes with highly desirable characteristics - to increase production. In this process, farmers can replace thousands of wild varieties and primitive cultivated varieties overnight with one high yeilding variety. If the displaced varieties are not collected and properly stored, they could be lost forever, India has already lost thousands of traditional varieties of wheat since the beginning of Green Revolution.

Simultaneously, environmental destruction caused by shifting cultivation, deforestation and dam building is destroying many wild varieties of even domesticated crop plants, trees and animals. Water pollutions affecting fish stocks and probably even genetic diversity. Overgrazing is affecting grass and free populations. However, no gene in this diverse nature can be considered useless and, thus, deserving to be driven to extinction. India possesses an extremely rich diversity of plant and animal life.

The Indian region is considered one of the 12 centres of genetic diversity in the world. The total Indian flora consists of about 45,000 species of which about 30,000 are non-flowering plants like algae and fungi which are usually inconspicuous and small. The flowering trees and plants of India consists of about 15,000 species but this number should not be taken as the final figure as several areas in the country remain unexplored and underexplored. From a very small area, silent Valley in Kerala, covering less than 9000 hectares, the Botanical

Survey of India recently discovered nine species new to science and a new genus of plants. About 7000 endemic species are found in India, that is they do not occur any where else in the world. Of these the Himalayas and the Khasi Hills in Meghalaya account for about 3000 endemic species, and the Deccan peninsula for about 200 species. The area of maximum diversity in India is northeastern India which is also the most threatened. The second major area of genetic diversity is the lower region of the Western Ghats around the Nilgiris. Over 1500 of the rare flowering plants are subjected to a lesser or greater degree of threatened extinction.

Just as India has benefitted from plant and animal genes from all over the world, India, too, has given to the rest of the world many valuable genes. Probably the most dramatic case is that of a single sample of wild rice, Oryza nivara collected by botanist S.D. Sharma from eastern U.P. in 1963. Grassy stunt virus is one of the four major diseases of rice plants. During the early 1970s, grassy stunt epidemics destroyed more than 116,000 hectares of rice in Indonesia, India, Sri Lanka, Vietnam and the Philippines.

At the International Rice Reserch Institute in Los Banos (Philippines), Gurdev Khush and K.C.Ling searched 6723 samples of cultivated and wild rice varieties looking for resistance to grassy stunt virus. Only one sample, that of Sharma's displayed the resistance they were looking for. Today, the genes from this sample are used in resistant varieties grown on 30 million hectares in India, Nepal, Bangladesh, China and Southeast Asia. The disease has almost ceased to exist in Indonesia. These genes have helped push average production per hectare to six tonnes and in some cases to 10 tonnes. Two or three crops have become common and rice production overall has doubled. In an interconnected world, Sharma's rice sample has helped India to repay a debt. Indonesian wild genes have helped India to set up a sugarcane industry. The limiting factor to commercial sugarcane production in India used to be the red rot diseases.

The resistance was acquired from the wild Indonesian cane Saccharum spontaneum, Indonesian S. spontaneum genes are now present in important commercial varieties of sugarcane used in such top sugar-producing countries as Cuba, Mexico, USA, Argentina, Brazil, Columbia, China, India, Indonesia, Pakistan, Philippines, Thailand and Australia. Today equally widespread are S. spontaneum clones from India, which are sources of resistance to gummoss and root rot. The list of valuable genes discovered in India is long. Large areas of the world, for instance, are covered by unproductive saline soils. Tolerance to salinity in rice has been tound in several traditional varieties from southern India and Sri Lanka. A wild melon from India has provided resistance to a form of mildew and has saved the melon crop in California, USA.

Among animals, too, many Indian breeds are known all over the world. The home tract of the famous sahiwal cattle is in the Sahiwal district of Pakistan. At the time of partition, only a few Sahiwal herds existed in India. A few more animals were brought by migrating Sikh and Hindu families. Today, the Sahiwal population in India consists of about to 1000-1100 breeding females. The breed is being used for crossbreeding in many countries. Sahiwal semen is being produced on a large scale at a farm in Queensland (Australia) for export to New Zealand, which, in turn, after contract matings by use of Sahiwal semen on Jersey and Friesian cows, exports crossbred heifers to Southeast Asian countries. The famous Murrah, Nili-Ravi and Surtibuffaloes are being used to upgrade buffaloes in many countries of the world. In Vietnam, a consignment of Murrah buffaloes is widely regarded as the most valuable gift from friendly India. The Jagarabadi has been successfully introduced into lands in which buffaloes did not exist untill recent times, for example, brazil and Trinidad. The Jamnapari goat breed of the Chambal ravines in Etah district(UP) has been used for improvement of native breeds in several countries. The famous English Anglo-Nubian breed, which has been extensively used for crossbreeding, is itself based on a cross of the Jamnapari.

Even germplasm of economically important insect needs to be preserved. The two insect livestock groups of major world importance are honey bees and silkworms. There are many strains of honey bee, and there is great variation within the strains for characteristics such as honey production and disease resistance. Though efforts to collect wild bee germplasm seem to be a long way off because of various technical reasons, the need is being increasingly felt. Work on breeding has already been done with the mulberry silkworm. In India, wild forms of the tasar silkworms (Antheraea species) are being used for improvement. Antheraea roylei from India and A pernyi from China have been crossed to produce the hybrid A, proylei. Silk from the hybrid is claimed to be the finest ever produced in India and the new species also out yields its parents by almost 170 percent in weight of silk and 94 percent in filament length. The hybrid is expected to play a major part in eforts to expand the tasar silk industry and provide rural employment.

In a recent lecture on genetic conservation, Dr. M.S. Swaminathan, one of India's leading agricultural scientists and Director-General of the International Rice Research Institute, drew attention to the need to conserve even micro-organisms. There has been an enormous growth of interest in biological nitrogen fixing organisms because of the ill-effects of chemical fertilisers on the environment. Among the major sources of biofertilisation are blue green algae and azolla. Azolla germplasm from all over the world is already being collected at the International Rice Research Institute.

Azolla is an aquatic fern capable of fixing atmospheric nitrogen through a symbiotic blue green algae living in the cavities of its leaves, it has long been used in China and Vietnam as green manure in rice fields and its use is growing in other Southeast Asia countries flow. Blue green algae (BGA) is being extensively promoted in India as a cheap source for fixing nitrogen in paddy fields. Unfortunately, there is no organised effort to collect BGA for genetic conservation purpose. Strains recorded during ecological studies are not kept by the scientists because maintenance of a culture collection is time consuming and expensive.

Scientists have achieved considerable success in selecting strains of Rhizobia nitrogen fixing bacteria which live in the roots of leguminous plants, and give them their nitrogen fixing abilities. This success indicates the possibility that superior strains may exist in other nitrogen fixing organisms, too. Today, the world's main food supply is determined by only 20 crops specis, and only eight forest species provide most of the world's wood and timber. Similarly, about 20 animal species provide most of the world's milk, wool, hide and draught power.

A much large number of plants to provide food, such as oil, vegetable, fruits and fodder, and meet other needs, such as spices, beverages, medicines, ornamentals and industrial raw materials. However, all these plants and animals constitute only a fraction of all the plant and animal species in the world. It is impossible to guess what this natural wealth will yield in the future. But the potential can be seen from the vast variety of medicines alone that plants have given to human kind. The Ayurvedic system of India gave to the world reserpine, a tranquiliser and a treatment for hypertension, anxiety and schizophrenia. Reserpine is a therapeutic agent extracted from a herb, Rauwolfia serpentina, which is now almost impossible to find in forest.

In India, at least 2500 plants are used for medicinal purposes. The roots of the berberry shrub found in the Himalayan foothills are an important source of berberin, a compound used widely in tropical Asia and Japan against diarrhoea. A single species, Strychnos nuxvomica, indicates the commercial value of drug producing plants. A source of the compounds strychnine and brucine, species earns India over Rs.50 lakh yearly in exports. India's total exports of plant-based chemical products, mainly medicines and pharmaceuticals, are worth over Rs.55 crore yearly, a trade 'which increased five-fold during the 1870's. It is distressing that overexploited medicinal plants face probably the greatest threat of extinction in India today. From an ethical point of view too, there are strong arguments against human beings driving any life form into extinction. But protecting the wonderful gene bank provided by nature is an issue on which both utilitarian and ethical perspectives coincide.

India is known especially for its rich genetic wealth of bamboo, fruit trees (mango, banana and citrus), vegetables like eggplant, akra and cucurbits (melons and gourds), grain legmes (mung beans urad or black gram and cowpeas), rice, pepper, cardamom, ginger, turmeric, sugarcane, jute, and various medicinal and aromatic plants. There is also a rich diversity in wheat, maize, millets, oilseeds and various other fruits, spices, vegetables, fodder and plantation crops. Plant species from India have not only influenced agricultural development in the region but also in many other parts of the world. Plants were taken from India to the ancient civilization of the Tigris, Euphrates and Nile river basins. Many Arab traders took plant materials to Meditarranean countries in the 8th-10th centuries AD. Buddhists took several Indian cultivated plants to Southeast Asian Countries. It is estimated that about 250 India species have wild plants of agricultural and horticultural importance and of these about 60 are rare or endemic. Though many wild varieties have neen collected and used in breeding programmes, only a fraction of the job has been done. India is also very rich in its natural wealth of forage legumes and grasses. About 2000 such species occur, of which tow-thirds are grasses.

The northeastern Himalayas are a veritable mine of valuable genes in many crops and fruit trees. Unfortunately because of shifting cultivating and widespread habitat destruction, this genetic wealth faces the threat or extinction. Dr. Bhag Singh of the National Bureau of Plant Genetic Resources (NBPGR), for instance, discovered *citrus indica,* an endangered species of citrus, in the Tura range of the Garo Hills, Meghalaya. This species is the most primitive of citrus species and perhaps the progenitor of cultivated materials. Citrus fruits are native to South Asia and the northeastern Himalayas hold a notably important position in the world with regard to citrus wealth. The Khasi and Garo tribes grow citrus trees in and around the vicinity of their houses.

A major crop in India and Asia, rice has attracted the maximum interest amongst indian scientists interested in plant genetic resources. The areas with maximum diversity in India are the northeastern tracts and the hilly regions of Orissa, Bihar, Madhya Pradesh and Andhra Pradesh, which are mostly inhabited by tribal population. Various local rice varieties have also been found in the malabar region and in the cold Himalayan tracts. A national survey estimated that although about 50,000 local rice varieties are still in cultivation, they are being rapidly lost, particularly in the irrigated tracts, where new high yielding varieties have been adopted, growing deforestation poses a serious threat to many wild varieties of rice. Tropical rain forest habitats have been important sources of disease and pest resistant varieties of rice.

The Silent Valley forest reserve area in Kerala, which was to have been inundated with the construction of a hydroelectric project, has been the Northeast are another invaluable source of rice germplasm in the form of the well-known Assam Rice Collection (ARC). Together with the threatened cultivated varieties, these endangered natural habitats should be a priority area for immediate collection of wild varieties. In a germplasm conservation programme, the wild species are as important as, or more important than, local cultivated varieties. A total of nearly 15,200 local rice varieties are being maintained at the national rice gene bank at the Central Rice Reserach Institute (CRRI) at Cuttack, and an estimated 35,000 local varieties are held as working collections at 40 rice reserch stations. Thus there is a total collection of 50,000 cultivars (Cultivated varieties), Assuming 50 percent of the collection to be duplicates, the national collection is now estimated at 25,000 with scope for further collection of another 25,000 varieties.

The genus Oryza (rice) consist of about 24 species, out of which only two are cultivated: O sativa (Asian rice) and O glaberrima (African rice). Collection of wild varieties has logged behind those of cultivated varieties. The group O, rufipogon, O, nivara and O, meridonalis, is the most closely related to the widely cultivated O. sativa rices and represents the progenitors or the cultivated rices. Although as genetic resources, their importance is paramount, unfortunately, they are fast disappearing. This group occurs in one of the densely populated areas of the world, where the habitat is being reclaimed for cultivation of crops, construction of roads and buildings. The cultivation of O.sativa rices is coming closer to the population of these wild species which are hereby losing their genetic identity by being "swamped" by the genes of O.sativa. A time bound collection programme for the wild rices in this group is thus urgently needed.

The group O.Collina, O.officinalis, O.minuta and O.malampuzhaensis are Asian rice species which occur in forests in partial shade, by the side running streams or swamps. With deforestation, this group is also endangered. The present collection of O.officinalis from South Asia is not representative of their natural distribution. It is important to collect these species before they disappear. In India, the CRRI is aided by 10 base centres at Simla, Almora, Jodhpur, Ranchi, Shillong, Cuttack, Hyderabad, Akola, Trichur and New Delhi. These centres collect not just germplasm for rice but also for awide range of other crops.

Each centre has been allocated a state or region in which it actively collects. In future collections of wild rice cultivated varieties,upland rainfed rices, flood-tolerance, pest and disease resistance, and aromatic and qualitycharacters will be emphasised. Collecting these wild species and peripheral populations under severe pressure may require specialised collectors and the creation of "gene sanctuaries" to protect threatened habitants. Accordign to Dr. M.S.Swaminathan, the collection of germplasm should not only involve professional scientists with expert knowledge of rice but also farmers, the general public and the defence services. In China, for example, each provincial academy mobilises farmers to send in seeds of traditional varieties. Although some samples will duplicate, this system lessens the work of the scientists and lets them concentrate on classifying the material and identifying duplicates and distinct types. As some of the high priority germplasm collection areas and end angered habitats are either sensitive border zones or inaccessible regions, defence personnel could also play an important role in collecting germplasm. Amongst the most vulnerable plant species in India are overexploited medicinal plants, ornamental plants, like orchids and botanical curiosities like pitcher plants. Some 1250 species of orchids occur in India, of which 300 are found in Meghalaya alone. About 20 species of orchids are becoming scarce in Meghalaya. Scientists from the Botanical Survey of India (BSI) recently made intensive studies on a group of beautiful orchids called Coelogyne, but failed to see even one plant of three species, namely, Calbolutea, C.assamica and C.treutleri in living condition. They have probably become extinct. Even their preserved specimens are not available in any herbarium. The Government of Meghalaya has recently established a "pitcher plant sanctuary" in the Khasi Hills. The Indian Orchidarium of Shillong has brought approximately 350 indigenous orchid speices into cultivation for conservation purposes. But much more needs to be done in this connection. For example, arhododendron sanctuary is urgently needed in Sikkim.

The loss of valuable plants is definitely a matter of great concern, though new attempts have been made as yet to collect their germplasm. Amongst the endangered medicinal plants are three orchids. Paphiopedilum druryi a native to kerala, which is endangered or extinct in the wild due to forest fires and excessive collecting; Dendrobium pauctiflorum found in West Bengal and Sikkim in areas open to tree felling which is also endangered, possibly extinct; and Diplomerial birsuta, restricted to a tiny population in West Bengal and is vulnerable to landsliders. A decrease in its number has been noted by botanists. The numbers of another orchid, Dendrobium nobile are seriously declining. This orchid occurs in the Himalayan regions of India and China and is the source of dendrobine which is still exported from China by the ton in a dehyarated from.

Several other medicinal plants are being seriosly threatended because of over exploitation. A corus calamus, known as vacha in India, is used in 51 different drug preparations. It used to be common in Manipur and the Nega Hills, but today it is so scarce that it has to be imported from overseas, Rauwolfia serpentina, the source of resperpine, known to Ayurveda for centuries, was once common and abundant in forest areas all over peninsular India. But after scientists isolated the alkaloid reserpine in 1952, a ruthless search was started all over the country, which came to a halt only when the plant almost disappeared from forest areas. Today it is extremely rare in nature and is mainly found in cultivated stands. Variations within the species may have been lost already.

Two other species that are extensively used in Ayurveda and are today endangered are Saussurea lappa, the source of kuth roots, and Dioscorea deltoidea, the source of steroidal hormone. Today these plants are listed in the Convention on International Trade in Endangered Species and cannot be exported without an export permit. Dioscorea deltoidea, much sought after by private agencies and pharmaceutical firms, has been continuously collected in India except perhaps in the inaccessible areas of the Himalayas. The roots yield cortisone, a steroidal hormone used in treating rheumatic diseases and ophthalmic disorders.

The need to preserve the medicinal plants of India is clearly vital and urgent. Maintenance of botanical/herbal gardens will not be a solution. Efforts will have to be made to conserve the germplasm of each important species of medicinal plants. The National Bureau of Plant Genetic Resources (NBPGR) has only made a beginning by collecting Opium Poppy, Syllium, Senna and Rauwolfia serpentina. The NBPGR has a major role to play in the protection of the country's vital genetic treasure. But the Indian council of Agricultural Research (ICAR), which supervises NBPGR, will have to be more effective in energising this organisation that it has been so far. Germplasm conservation is a complex task.

Clear priorities have to be set in collecting germplasm. There should be an understanding of how existing collections or duplicates should be weeded out. Collected seeds should be stored carefully. They must be properly characterised for their attributes and then made available to breeders for use. Finally, there should be adequate facilities for multiplication of seeds while duplicate samples should be kept as an insurance.

Although the NBPGR was set up in 1976, it has still not built a proper facility for collected germplasm. Funds were allocated for the construction of storage facility in 1978 but it took three years just to choose the site. It has taken a further three years to finalise architectural details, with central public works department officials anxious to visit Europe to look at existing storage systems. There is still no firm date by which the storage facility will be built. In such circumstances seeds have to be sown every year and recollected. One can only guess how well the specimens have been kept. It is quite possible that many samples have been lost or are mixed up.

The state of the rice collection at the Central Rice Research Institute should also be a matter of public concern CRRI too does not have a long-term storage facility. In the humid climate of Orissa, the viability of rice seeds is generally not more than six months. Consequently, many scientists lack confidence in NBPGR and prefer to keep their collections with themselves. The most extensive collection of rice from the Chattisgarh region was made by Dr. R.H. Richharla. Beginning in 1971, while workings as the agricultural adviser to the government of MP, Richria collected, with single minded devotion, some 18,000 samples of rice.

Though many of these may be duplicates, Richaria's vast collection become the nucleus for the breeding activities of the MP Rice Research institute (MPRRI),which began in 1976 and was later merged with J.N. Agricultural University. Richaria, in fact, argued that it is the higher-yielding among these indegenous varieties which need few inputs, many indegenous varieties will outyield the so-called high-yielding imported varieties. The work of collecting rice varieties in Chattisgarh has almost come to a halt since MPRRI was close down in 1978 allegedly under World bank pressure Doubts have been expressed about the care with which Richharia's collection has been kept.

India has a rich diversity of animal life which is threatened by a number of factors. Even as scientists discover new breeds like the Siri cattle in Sikkim, and although many species are nowhere near extinction, they are undergoing a process of genetic erosion, Urban centres attract cows and buffaloes with the highest milk yield from the rural areas. But from there they go to the slaughter-houses, after completion of location. Thus, high producing animals (genotypes) are being regularly removed from the breeding tracts, leaving inferior genotypes for breeding. P.N. Bhat, one of India's leading animal geneticists, believes that the genetic quality of the famous Murrah breed has deteriorated. The maximum recorded milk yield in a location for a Murrah buffalo was obtained in 1932 with one female yielding about 4000 kg. In a lactation of 310 days, it is difficult now to find buffaloes recording better that 2500 kg per lactation. It has been estimated that between 1921 and 1979 the loss in average lactational yield of good producers due to the genetic erosion caused by slaughter-houses and other factors has been of the order of 1000kg.

India has 26 breeds of cattle and seven of buffaloes, together with a number of nondescripts, which account for about 75 per cent of the total cattle population. The improved and specialised breeds are generally found in the northwestern and western dry areas. Indiscdriminate crossbreeding with exotic breeds poses a serious threat to indigenous breeds. Already crossbreeding with temperate breeds has made in roads into good drought breeds. Bhat argues that the need for crossbreeding in their breeding tracts needs to be evaluated and stopped if erosion is large, so that the local breed may not be last.

Some of the finest milk breeds are facing constant erosion, says Bhat, both because of indiscriminate crossbreeding and neglect. There have been no practical breeding programmes for genetic improvement of the indigenous breeds. These breeds are not in as good a genetic condition as they were a few centuries ago.

Mass artificial insemination and import of exotic stock for cross breeding has now made it impossible even to locate geneticaly superior animals of 'Sahiwal, Red Sindhi and Tharparker breeds in India. "Purespecimens of rathibreed are fast becoming difficult to find; similarly high producing kankrej and Gir animals are becoming rarer and rarer," laments Bhat. Among cattle, the good milk breeds like Sahiwal and Red Sindhi and dual-purpose breeds like Tharparker and Rathiare in acute danger of being lost in their pure form India.

Bhat himself believes that livestock breeds which can sustain and produce on low agricultural inputs should be identified and selected in animal rescarch programmes in tropical countries. There is little point in introducing animals which need high inputs to produce efficiently in conditions of poverty. Apart from the humped cattle, there are in the northestern Himalayas a number of wild, semi-wild and domesticated humpless cattle. Economically the most important is the mithum (Bos frontalis), raised primarily for meat, sacrifice and bride price. The mithun is believed to have arisen from the cross of gaur, the wild cattle of the indo-Malaysion region, with domestic cows. The 1972 census revealed 121,000 mithuns in Arunachal pradesh, Nagaland, Tripura and Manipur. The mithun lives mainly in semi-wild conditions at an altitude of 600-1600 metres.

Among the lesser known cattle, but extremely important for Himalayan traders, is the yak, or grunting ox (Bos Poephagus graunieus Linn). In the wild it inhabits the coldest and most desolate tracts of the Himalayas, from the east to the west, with elevations ranging from 4000-6000 metres. Without the yak, travel and trade in these parts would be as impossible as without the proverbial camel in the desert. According to the 1972 census, there were 43,000 yaks in India. The yak interbreeds with many types of cattle: bison banteng, goyal, zebu and European cattle, and has been so interbred for centuries. No interbreeding with buffaloes has been recorded so far in India. Yak bulls are mostly used for crossbreeding hill cows. The hybrids are hardier than the yak and are capable of bearing heavier loads, but are inferior in staying power. Hybrids with cattle are now twice the number of purebred yaks. Since there has been no organised effort to preserve this useful animal, its population has declined by almost 25 percent over the last 10 years.

Though sheep are very important in certain regions of India, little work has been undertaken to identify whether any valuable Indian sheep breeds face the danger of extinction. It may take as much as 25 years before a clear picture of the available genetic material in sheep emerges. Scientists argue that evaluation of indigenous sheep breed for corssbreeding and their conservation should go hand in hand. It should be mandatory for all sheep development programmes undertaken by state governments to conserve nucleus flocks of breeds used for crossbreeding in their natural habitat.

Like sheep, goats are also important to the Indian rural economy, and as one goat expert puts it, despite regular attempts by soil conservationsists to exterminate this animal, its population has been increasing. The origin of Indian goat breeds is not clearly known. They are believed to have been derived from wild goats which inhabited the Asian mountains in antiquity, and to have been domesticated around 7th century BC much earlier than cattle. There are at present about 17 distinct goat breeds in India, of which several like the Jamnapaari are extremely useful. The dwarf goat breeds, such as Black Bengala, Banbari, maalabari and Assam Hill are famous for high prolificacy (multiple births) and early sexual maturity.

The pashmine goats of Ladakh produce the finest quality of Pashmina fibre (cashmere) in the world. But though Pashmine goats support a major cottage industry in the Ksahmir valley, little is known about them. Research on goats is of recent origin and specific strains are now being developed for milk, meat and fibre, Crossing of pashming goats with angoramight make possible a goat with mohair and poshmina. Gaddi, a hill goat, produces hair used for ropes. Crossbreeding of these goats with angoras yields mohair.

India had only about seven million pigs in 1976 but they are an important source of meat espcially for enonomically disadvantaged groups like Harijans. Traditionally indigenous pigs produced bristies and meat. In fact, meat was only considered a by-product, until the advent of nylon which replaced pig bristles. There has been no serious attempt to catalogue the various breeds or strains of domesticated pigs in India, and previously unknown varieties are some times found for example, P.N. Bhat recently located a number of strains of dwarf pigs in the north eastern region. Indigenus pigs can survive under extremely adverse conditions of malnutrition and health: They are reportedly more resistant to disease than exotic breeds. Pigs are receiving greater attention under government programmes and indigenous pigs crossed with exotic breeds are becoming more popular particularly in central India. Their higher growth potential makes them attractive, but given the adverse circumstances in which they are maintained, the survivability of corssbreeds is low, and therefore their advantage is marginal. Because of extensive crossing with the Large White, the indigenous pig in the central region is becoming extinct. Indigenous breeds need to be properly documented and their performance tested so that proper breeding strategies can be decided. Besides domesticated species, there are several wild ones, such as the common Indian wild boar. The present day pigs have evolved as a result of gradual domestication of wild pigs.

India is one of most important countries as for as camel husbandry is concerned its population of 1.1 million camels constitutes about seven percent of the world total. Camels are found mainly in Rajasthan, Haryana and Punjab, the Arabian one-humped camel being the only species. There are three types of camels found in India: the desert camel, the riverine camel and the hill camel. Bikaneris the most important desert camel breed in India. Nearly 50 percent of all Indian camels belong to it. Riverine camels are found in districts of Uttar Pradesh and Punjab, where there is a good water supply. They carry heavy loads and are slow in movement with a speed of 3 km/hour. The desert camel can walk 190 km overnight at a speed of 15 km/ hour. The hill camel is more compact than the camels in the plains. The National Commission on Agriculture (NCA) had expressed concern over the changes taking place in the traditional system of camel breeding practised over centuries in Rajasthan. Under the traditional system, camel breeding was undertaken by professional breeders under the patronage of princely rulers and large landlords, who maintained large herds of camel. Excellent camelherds were owned by Bikaner, grazing of camels with no tax levied on their grazing.

Drinking water from all the wells in the state was allowed free for the animals, Good stud camels called "Mehias" were branded and let loose by individuals intrested in camel improvement. These stud camel would move in herds in winter and during the offbreeding season they were free to graze anywhere, without fear of impoundment even if they damaged standing crop. The traditional camel breeders, with their keen observation, used to breed camels with a definite objective. The camel breeders in Rajasthan evolved distinct types of camels like the Bikaneri and Jaisalmeri. Today, however, the conditions favouring breeding of camels are fast disappearing. Grazing lands have diminished at a rapid rate, having gone under the plough.

As a result most camel breeders are now leading a pecuniary existence with small herds of camels in poor condition. Under these adverse conditions the fine art of breeding has disappeared, reduced to a mere multiplication of stock. "It may not be long before these camel breeders will go completely out of business", warned the NCA. The NCA was not even sure whether there is a need to make a special endeavour to retain the distinct types of camels that were evolved in the past. "Breeding of animals for genetic improvement without ensuring adequate nutrition is meaningless", it arugued. The construction of roads in desert areas is making the animal itself redundant.

In the past camels were prized most for their speed of travel and capacity to traverse long distances at a stretch in the desert. In increasing numbers, camels are today being employed for agricultural operations. They played an extremely important role in the construction of the Rajasthan Canal. The survival of the animal, and its various types, is now dependent on whether adequate grazing lands will ever be found for it.

India also has about 0.94 million horses and ponies, 0.075 million mules and 0.99 million donkeys. Uttar Pradesh has the greatest number of horses and mules, and Rajasthan donkeys. Horse breeding has been practised in India for a long time. There are six important breeds of Indian horses and ponies, which include the Kathiawan, one of India's best breeds which is found in Rajasthan and Gujarat: Marwari, which is closely associated with the great feats of bravery attributed to the rathor Kings of Rajputana: Bhutia, which is bred along the Tibetan border and sub-Himalayan tract from punjab to Darjeeling: Manipuri, which has been bred in Manipur and Assam for many centuries: the Spitipony of the Spito Valley: and the Zanskari horses of leh in Ladakh. Donkeys are of two types: The small Grey, which is commonly found in most parts of the country and the large white which is found in the Kutch.

Unlike the camel, the donkey has never received much attention. there has been no organised donkey breeding in India even though it is an important work animal. There are no textbooks or monographs devoted to the donkey, no research on its physiology, productivity or work ability, no selection experiments to increase its performance, even though next to the camel, it is the best farm animal adapted to a hot, dry environment.

Chickens constitute over 90 percent of the total poultry population of India, followed by ducks, six per cent turkeys and geese, 2 per cent, India and the neighbouring countries have been referred to as the original home of the Red Jungle Fowl (*Gallus gallus*), from which present day domestic birds are believed to have descended. The Aseel or Malay Fowlis reported to have given rise to all the breeds of poultry. There is evidence to show that these birds, some 2000 years ago, moved through the Middle East to Europe and gave rise to the present day European breeds.

Eighteen types of indigenous breeds are on record. But in number, they are much smaller than exotic breeds. The former number about 50 million as compared to about 100 million of exotic breeds (including crosses). Several indigenous breeds are facing extinction. The original Aseel, a native of Andhra Pradesh, is a medium sized, round-bodied and well-built bird.

However, large specimens have also been bred and developed by the promoters of cock-fighting. The bird is poor layer but is known for its delicious meat. With the ban on cock-fighting, the Aseel is losing much of its importance, and as a result pure specimens are becoming rare. The Ghagus is a large sixed bird, which is usually found running with the nomads of Karnataka and Andhra Pradesh. It is a hardy breed and is supposed to have resistance to diseases and parasites. Due to neglect, it is gradually becoming extinct.

Information on ducks and other birds is extremely limited, there appears to have been no evaluation of indigenous ducks. At a recent international meeting a scientist expressed concern over government plans to establish breeding farms for exotic Khaki Campbell ducks, both for intensive rearing of these ducks and for upgrading local ducks. There is no specific information as yet that under normal husbandry conditions of limited feed supply and no-existent disease control, the exotics are better than the indigenous birds.

Animal conservation is still a new area even for scientists. With the failure of several animal development programmes in which exotics have been introduced on a larger scale or crossed indiscriminately with local breeds, the importance of environmental adaptation using local breeds is being recognised. Hence, the importance of conservation of indigenous breeds to support more appropriate breeding programmes is also acknowledged.

Conservation of animal genes is neither easy nor cheap. It can be done either by maintaining a living herd or flock of animals that need to be conserved, by storage of frozen semen from a representative sample of males or by storage of frozen embryos.

In India the need to conserve animal genetic resources is being recognised. Several institutes concerned with preserving animal genetic resources have been set up. They are the National Bureau of Animal Genetic Resources (NBAGR), a National Centre for studies on the Mithun in Arunachal Pradesh, a National Institute for Research on Goats in Mathura district, Uttar Pradesh and a breeding farm for camels of Bikaner. An enormous amount of work is needed to document and evaluate indigenous breeds of birds and animals and then based on this information, urgent action taken to conserve breeds.

Though India has a rich variety of fish fauna only a few species have ever been cultivated. The most important among these are the carps including Catla catla, Labeo rohlta, Cirrhinus mrigala, labeo calbasu and a few others. Several exotic species have also been introduced into India (the Chinese carps, tilapia, trouts, gourami, etc). The natural habitats of the cultivated species are the major rivers where they breed naturally. Under aquaculture, however, though they mature in ponds, they normally do not breed there. Fish seed, therefore, has to be collected from the rivers for aquaculture. As it is being practised today, this will have a detrimental effect on fish and may leave a genetically deprecated stock in the natural habitat.

Since fish species inhabit different ecological systems in geographically and climatically in isolated areas, it is possible some of these species could have evolved into definite strains having various advantageous characteristics. Unfortunately, no comprehensive study has been made so far even on these few cultivated fish.

Intensive aquaculture through genetic selection and selective breeding result in the species becoming more inbred and with less genetic diversity. Though aquaculture work in India is still limited, original diversity could be eroded in future. Thus to maintain the genetic diversity for present and future requirements, conservation of genetic resources in fish is essential. Several dams and canals are being built to harness the country's water resources. They can have serious implications for fish genetic resources by permitting inter mixing and breeding of different strains and species which hitherto enjoyed isolation and preservation in different geographical and ecological boundaries.

Transmittance of undersirable species to areas where they are not liked or which may compete for food space and breeding with local varieties; introduction of fish predators and various other organisms; introduction of fish parasites and dangerous pathogens to areas where they were not known before; and spreading of undesirable and harmful aquatic weeds are some of the more obvious possibilities. In the case of crop genetic resources, after exploration, varieties are conserved in a central pool. Thus at any given time the required strain is available from a central place. But maintenance of fish at a central pool for a number of years can be hazardous as there can be complete loss of a precious variety through disease and other unfavourable factors. Furthermore, maintenance of certain strains over decades in ponds and repeated breeding will result in partial or complete domestication of the species. This is contrary to the idea of conserving genetic resource.

The best way is to protect these varieties in their natural habitat. This needs legislation to declare certain natural habitats protected and free from fishing; preventing pollution of the protected zone; preventing construction of dams and weirs in these protected zones or if dams and weirs are constructed, suitable and drainage system; protecting fish breeding areas in natural habitats; and, restricting large-scale exploitation of fish seed from riverine sources. The Indian Council of Agricultural Research (ICAR) has recently established a working party on conservation of genetic resources of fish to consider all aspects of this problem. But this constitutes little more than a beginning of the understanding of the problem.

Large areas of forests are in imminent danger of disappearing. Although there is little risk of entire species of trees becoming extinct, populations at the limit of the species range or in isolated block of forests are definitely vulnerable. Often these small populations have a high frequency of valuable genes-for drought resistance, for example-which could be used in future tree breeding programmes. Tree genetic resources have been neglected in the past. With a growing interest interest in forestry across the world now, it has becmoe essential to preserve natural wealth in trees. South Korea, which has carried out the most extensive afforestation in Asia, has set up on entire research institute devoted to tree genetic resources and development of suitable trees.

Tree populations are most threatened in arid and semi-arid areas where an intense pressure exists because of the demand for firewood. Developments in commercial forestry have put imense pressure on specific trees which have acquired an economic value. The Dang forests in Gujarat, for instance, are a rich reservior of a number of plant species which were not considered of economic value about 20 years ago.

Good examples of these species are kakad (Garuga pinnata), which is used for pulp and packing cases, tanch (Ougeninia oojeinesis) used for making agricultural implements, haldu (Adina cordifolia) and kalam (Mitragyna parvifolia) used for making agricultural implements, haldu (Adina cordifolia) and kalam (Mitragyna parvifolia) used for making bobbins and furniture, and modad (Lannea grandis) and kudi (Wrightia tinotoria) used for making toys. It is important that tree species in this area, as in many other forest areas of the country, be preserved.

Equally, several village woodlots, with trees representing well-adapted genetic material, are being steadily depleted. FAO wants to give priority to collecting the diversity in tree species suitable for firewood and fodder production in dry areas. Arunachal Pradesh is another area where the richness in tree resources is being steadily threatened. As two scientists of the Forest Research Institute of Dehradun put it: "In spite of the richness of the flora, the area, to this date, remains a terra incognita and without a comprehensive published flora".

A paper published in "The Indian Forester" in 1980 listed abut 20 species facing danger of extinction or depletion. This list includes Phyllostachys assamica, a bamboo used for making huts, Gymnociadus assamicus, a small tree which grows up to 20 metres high and whose fleshy pods are used for washing hair as a substitute for soap, and the spectacular Rhododendron dolhousiae which has cream colour flowers with purple bands, R. dahousiae is such a beautiful plant that it was awarded a first 'Class Certificate by the Royal Horticultural Society in the UK. It is now facing extinction owing to massive road building and developmental activities in Arunachal Pradesh.

Tree genetic resources can be conserved either in situ or ex situ. Conserving in situ, that is, preserving the entire ecosystem together with its trees is definitely more desirable but often extremely difficult and expensive. When effective in situ conservation not possible, trees have to be conserved in artificial stands established ex situ in their country of origin or elsewhere. Conservation of tree genetic resources has been practised in India through the system of "preservation plots". The first preservation plot was set up in the sal forests of Bihar in 1906. Some 188 plots have been established since then to protect mainly commercial species like Shorea robusta (sal), Tectona grandis (teak), and various conifers. Most of these plots cover natural forest types and range from a few hectares to 4000 ha. Preservation plots in plantations are small. The coverage of these plots is also patchy. Many forest types like southern tropical forest, semi-evergreen forests, and tropical thorn forest, need particular attention. A beginning is being made in the preservation of trees important for social forestry, Leguminous shrubs, which can enrich soil fertility through biological nitrogen fixation and at the same time provide food, fuel and fodder, are particularly worthy of greater attention.

Leucaena is already receiving worldwide attention. Several international conferences have recommended the collection and conservation of casuarina and neem germplasm. Two conferences held on neem in West Germany in recent years have pointed out that the potential for improvement of the neem tree in yield and quality characteristics is considerable. India is participating in an international project to collect tree species used in arid and semi-arid areas. According to the list of priority species that has been prepared for this project, first prority will be given in India to three indigenous species (Acacia nilotica, Acacia senegal and Pros popis cineraria) and several exotic species of acacia, prosopis eucalyptus, and leucaena.

Among the second priority indigenous species are Albizza lebbek, pongamia pinnaka (karani) and Azadirachta indica (neem). Among exotics given second priority is Prosopis juliflora, which was introduced into India nearly a century ago but which is now widespread and shows considerable variation of form. Indigenous species listed for future consideration are hardwickia binata, Kingiodendron pinnatum and Dendrocalamus strictus (bamboo).

Born completely blind, the Gangetic dolphin (okatabusta gabgetuca), remains sightless for the rest of its life. Nevertheless it does get by perfectly well, having adapted to a sightless existence in its murky, river habitat. Not to be confused with marine dolphins (Dephindae), the fresh water dolphins (Platanistidae) have lost all contact with the sea. They search for fish crustaceans and other such organism by sweeping the river bed with their sensitive muzzles. Fresh water dolphins must come up for air every one or two minutes. During the day the dolphins perfer to stay in deep water, coming out at nightfall to feed in the shallows. Hydroelectric projects and the large fish nets of fisherfold pose considerable problems for fresh water dolphins but in recent years positive steps have been taken to identify and protect some of their known habitats in the Brahmaputra, Indus and Ganges river.

Civets are cat-like creatures with elongated bodies that resemble those of weasels. Solitary and nocturnal, these carnivorous creatures spend the day in hiding, coming to hunt small birds, mammals or reptiles at night. To accommodate their 32 to 40 teeth civets have evolved a longer snout which further differentiates them from the cat family.

In India, Sri Lanka and even further east lives a civet that has come to be known as the 'today cat'. This creature, the Plam, civet (paradoxurus hermaphorditus) has a tail as long as its body and derives its name from its odd preference for fermented toddy. It often lives close to human habitation and, as many have discovered to their unpleasant cost, can give off an extremely foul, clinging odour form its anal glands if disturbed. The palm civet is less carnivorus than its cousins and has specially adapted claws suitable for climbing the straight palm trunks.

An ant lion larva digs a steep tunnel in loose earth and lies in waiting in the opit thus created with only its deadly pincers and head showing above the surface. A small insect which wanders too close, slides down the rim and cannot recover its balance as its legs get no purchase from the crumbly slopes. The ant lion larva shoots up a spray of sand, which hastens the hapless victim towards its end. Having fed heartily on the digestible insides of its prey the tiny predators spits out the husk and patiently waits for the next victim.

The Indian Pangolin (manis crassicadudata) has evolved a scaly body covering, fabricated out of modified hair. As protection against predators, it trolls itself into a ball and raises its sharp scales. These anteaters are mainly terrestrial, have no external ear openings and can seal their nostrils to protect themselves from ant bites. Their long, extensile tongues are coated with a sticky secretion which allows them to capture ants, termites or larvae by probing deep into the labyrinathes of termite mounds which they rip apart with their strong claws. Pangolins have no teeth as their diet demands no chewing. A young pangolin will perch on its mother's back, just above the base of her tail, when it needs to be transported from one place to another.

The great Indian hornbill (Buceros bicornis) is large, ungainly bird with strange nesting habits, the female seals herself into a hollow in a tree where upon her feathers soon moult. The male dutifully brings her food for the entire period during which her eggs are incubated: feeding her through a narrow slit in the "prison" nest. When her young grow and their appetite increases, the female breakes down her prison doors to help the male obtain food. The young new rebuild the damaged portion of their home to seal themselves in. The hornbill's nesting behaviour serves to protect defenceless young efficiently.

4 Microbial Biodiversity

Roots, stems, leaves, flowers and fruits of plants all provide suitable habitats for microbial populations. For example, plant roots provide nutrients for microbial Growth and directly influence the composition and density of the soil microbial community. Whereas the microbial ecology and diversity of the rhizosphere (root system) has received a great deal of attention, the phyllosphere has been described as an ecologically neglected milieu. Some more attention has been paid to the microbial ecology of the aerosphere. Chen et al. investigated the fungal flora on twigs and fruits of chestnut in many areas of China. Results showed that many organs of chestnut trees were associated with microorganisms.

Among the fungi, the most common genera were *Phoma, Dothiorella, Sphaeropsis, Gloeosporium, Cytospora, Alternaria, Rhizctonia, Coniothyrium, Phomopsis and Fusarium,* although the type and abundance of the fungi differed on different organs in different areas. Fungal diversity decreased in the following order: Southwest China, the Yangtze River catchment, the Huaihe River catchment and Yanshan Mountain area.

The study also revealed that the ratio of parasites to saprophytes changed as the chestnut fruits developed. Saprophytes and weak parasites dominated at the early stages of growth whereas the population of parasites increased significantly as the fruits matured. Endophytic microorganisms are defined as all living microorganisms (saprophyte, parasite and symbiont) within the, plant body. Numerous microorganisms occupy cortical pores calyx tube, and core and vascular system. For instance in studies of the microflora of vascular systems in cotton plants, Lu found that the main fungi were five *Fusarium* species, three *Chaetomium* species, *Trichothecium* sp., *Trichoderma* sp., *Alternaria* sp., two non-spore forming species and some yeasts. The predominant bacteria were *Bacillus, Erwinia, Xanthomonas, Corynbacterium* and *Chromobacterium.*

Besides the species diversity of microorganisms (viruses, eubacteria, archaebacteria and eukaryotes), we can distinguish various forms of functional diversity of microorganisms, such as metabolic diversity (aerobic, anaerobic or facultative and phototrophic, chemotrophic, autotrophic or heterotrophic) and diversity of interspecific relationships (commensalism, synergism, mutualism, amensalism, competition or parasitism).

In the field of agriculture, the following functional characteristics of microorganisms deserve particular attention: nutrient supply and growth promotion for the plant ("biofertilization"), antagonism, competition, and mycoparasitism relative to plant pathogens ("biocontrol"), and degradation of pollutants in the environment (biodegradation").Among the nutrient suppliers, nitrogen-fixing bacteria are probably the most important group. They may be free-living (e.g., *Klebsiella pneumoniae)* or associated with the roots of either leguminous plants (e.g., *Rhizobium meliloti)* or monocots *Azospirillum brazilense)*. It has been shown that the effect of nitrogen fixation is most significant for leguminous plants whereas the benefits to non-legumes may be small. Indeed, the beneficial effect of *Azospirillum* may derive not from its nitrogen fixing properties, but from the stimulating effect on root development. Several bacteria may enhance mineral uptake by the plant by solubilizing soil phosphate.

In addition to nitrogen-fixers, also these bacteria may have considerable potential as bio-fertilizers. Many microorganisms are known to produce plant growth regulators (PGRs), including auxins, gibberellins, cytokinins and ethylene. Arshad and Frankenberger have summarized PGR-producing microorganisms, their soil in, it responses to PGRs.

Many microorganisms produce antimicrobial compounds which may act on plant pathogens by, inducing fungistatic, inhibition of germination, lysis of fungal mycelia, or by exerting fungicidal effects. For example, fluorescent pseudomonads produce siderophores which have both fungistatic and bacteriostatic effects by depriving the pathogen of iron, a limiting essential nutrient.

Chitinase produced by some bacteria lyses the hyphal walls of a Fusarium species. Volatile compounds, such as ammonia and HCN produced in the rhizosphere are also harmful to certain pathogens. Phytopathogenic bacteria generally occupy a particular ecological niche, but several beneficial bacteria compete with them for habitat and nutrients.

For instance, Pseudomonads and *Bacillus* spp. catabolize diverse nutrients in the root zone or on leaves and have short generation times. They are therefore logical candidates for biological control of phytopathogens.Some fungi can parasitize on other living microorganisms for nutrients. Perhaps the best-known mycoparasite is the fungus *Trichoderma,* hyphae of which may penetrate resting structures or parasitize hyphae of other microorganisms. This characteristic is of interest for biocontrol of plaint diseases. Because of the intense activity and growth of microorganisms, they have considerable potential to enhance the biodegradation of organic contaminants *present in* the environment, such as pesticides, herbicides, raw fuel oil and industrial pollutants. Microorganisms could be used for degrading contaminants both in and on plants and in the larger environment.

The crop plant and the microorganisms living on and in it constitute a holistic system with a complex equilibrium based on the interactions among the various microorganisms and between microorganisms and crop plant. Changes in this equilibrium, particularly alterations in the interaction between pathogens and other microorganisms, may affect plant disease occurrence and plant growth.

For instance Kloepper reported that most plant growth-promoting rhizobacteria (PGPR) strains that have been investigated to date appear to enhance plant growth indirectly by reducing deleterious rhizobacteria (DRB) populations. The relationships between plant and microorganisms are the result of coevolution: Host plant evolution is affected by pathogen stress, while pathogen evolution is influenced by host plants.

The final result is a stable balance, resistance to crop disease and pathogenicity of plant pathogens tend towards moderation. In the original habitat of the crop plant, plant and microorganisms (including pathogens) lived together for lone periods of time and as a consequence the plant-microorganism system is very complicated and steady.

Because of the diversity of microbial species arid function, microbial products are widely used in agricultural production, especially for plant disease control. The use of microbial products has the following advantages over conventional chemicals:

— They are considered safer than many of the chemicals now in use.

— They do not accumulate in the food chain.

— Self-replication of microbes circumvents the need for repeated application.

— Target organisms seldom develop resistance as is the case when chemical agents are used.

— Properly-developed biocontrol agents are not considered harmful to ecological processes and the environment.

At world level, a lot of microbial products have been developed for agricultural purposes and are being rapidly commercialised. For instance, the first US product consisting of a PGPR strain. QUANTUM *(Bacillus subtilus),* is readily available at the commercial level. The concept of microecotechnology, which covers biological control of plant diseases and pests, probiotics of animals and human, and bioremediation of environmental contaminants, has been well-received by many scientists in China. Indeed, microbial products are being extensively studied, continuously developed, and widely applied in China.

Microbial products are generally divided into two categories: pure culture microorganisms, and organic soil amendments or conditioners with living microorganisms. In this section, two cases of the use of microbial products in Chinese agriculture are discussed with an emphasis on their role in preserving microbial balance, promotion of crop growth, suppression of diseases and removal of contaminants.

Research on applying live microorganisms to crop production is rapidly developing in China. Most of the research involves the utilization of *Pseudomonas* spp. and Bacillus and undoubtedly one of the most successful examples is yield-increasing bacteria (YIB), YIB is a commercial name for a microbial agent made up of different strains of *Bacillus spp.,* mainly B. *cerues,* and *B. subtilis,* as well as *B. firmus, 8. lichenformis, B. coagulans, B. brevis,* and *B. sphaericits.* A group of Chinese phytopathologists led by Yanxi Chen from Beijing, Agricultural University had long studied many plant, diseases, such as black rot disease *(Ceratocystis fimbriatia)* of sweet potato and wilt diseases (*Fusarium*) of cotton.

On the basis of their experience, they introduced a concept called the natural ecosystem of the plant, which expressed the idea that an individual plant constitutes a microbial ecosystem, and developed a microbial agent called "yield-increasing bacteria" to improve the conditions of plant growth. The beneficial effects of YIB were first demonstrated in 1986, According to the Ministry of Agriculture, in 1991 YIB was applied to more than 50 types of crops covering 52 million ha of cropland in 30 provinces and municipalities in China. It was reported that the average yield of crops had increased by 10% or even more following the application of YIB to the field crops.

The screening process for YIB has been based on the following considerations: (1) the strain should be capable of adaptation and colonization in the target plant; (2) the strain should be strongly competitive with the indigenous microorganisms; (3) the strain should be beneficial to plant health and growth. The screening process for YIB consisted of the following steps. First, strains were isolated from the surface or from within the tissues of the plants grown in the field.

Then, YIB strains were selected by bioassay tests conducted in greenhouses rather than by traditional screening methods (Petri dish antagonistic test). The parameters for selection included percentage of emergence, final stand, root and shoot fresh weight, number of tillers, and disease resistance or suppression. Finally, strains were tested by replicated field trials. The microbial agent (YIB) is applied by either dressing seeds, spraying leaves of crops or soaking roots of crops in spore suspension. The application is generally more effective at the early stages of plant growth. Beneficial effects on the crop include earlier seed germination, larger root system, and greener leaves.

The application of YIB reduces the severity of some diseases, such as sheath blight on rice caused by *Rhizoctonia solani*, take-all on wheat caused by *Gaeumannomyces graminis* var. *tritici*, wheat scab caused by *Gibberella zeal*, sharp eye spot on wheat caused by *Rhizoctonia cerealis*, black rot on sweet potato caused by *Ceratocystis fimbriata* downy mildew on cucumber caused by *Pseudoperonospora cubensis*, bacterial wilt of sweet potato caused by *Pseudomonas solanacearum*, leaf rust on wheat caused by *Puccinia recondita* f. sp. *tritici*, and root rot on soybean caused by *Fusarium oxysporium*.The mechanism of action of YIB has been subject of considerable research. Studies indicate that YIB strains can colonize, migrate and multiply both inside and outside plant tissues.

Twenty-four to 48 hours after germination of rape seeds inoculated with antibiotic-resistant mutant of YIB strain 83-10, up to 10^4-10^5 cfu could be recovered from individual roots. The bacteria, inoculated in form of strains 83-10 or A-47 on the seeds of cabbage, maize and wheat, were found to be able to migrate to and colonize also the stems and leaves reaching a population of 10^2-10^8

cuf per leaf. Yu et al. sprayed YIB strain A-47 on the leaf surface of wheat and the number of inoculant multiplied rapidly after spraying, reaching very high numbers around the stomata, wounds and sunken sites according to scan electronic microscope observations, antibiotic double marker test (rifampicin, streptomycin resistance), and leaf surface replica test. The spore-forming *Bacillus* spp. have metabolically predominance for long periods and occupy ecological niches. They compete with pathogens for space and nutritional resources and have the ability to suppress, pathogen activity.

The population of DRB strains Bacillus 112 and 114 greatly diminished on the rhizoplane of rapeseed after seed-soaking with YIB strain 83-10 suspension. Liu indicated that application of YIB strain could eliminate the effect of some pseudomonads (DRB) in the rhizosphere of turnip.

By spraying the suspension onto wheat leaves in the field and then isolating leaf samples weekly, the average numbers of *Cladosporium* spp. and Al*ternaria spp.* on the surface of the leaves was reduced by 53.9% and 66.4%, respectively, compared to non-treated control samples. Some strains of YIB can produce antibiotics and inhibitory substances.

A liquid culture of YIB strain A-47 inhibited spore germination in *Puccinia striiformis* f. sp. *tritici,* and *Cladosporium spp.* by 71.94% and 87.32% respectively. The YIB culture also inhibited germ tube growth. Thus, YIB may reduce the incidence of infection by these fungi. Some of the YIB strains can secrete concentrated amounts of superoxide dismutase (SOD), an enzyme related to resistance, disease and aging stress. The SOD activity of rape (*Brassica napus*) tissue was found to increase after rape was treated with YIB filtrate.

The activity of peroxidase, which is related to disease resistance, also increased after treating wheat plants with YIB. Enzymatic analysis showed that YIB strains also contain other enzymes, such as amylase, proteinase, and alcohol dehydrogenase. The culture filtrate of YIB contains phytohormones, such as indole acetic acid (IAA), zeatin and gibberellin (G3). Peng and Xiong showed that YIB strain grown on[14] C-glucose medium could secrete IAA, G3 and zeatin.

Anaerobic digestion is an effective method for treating many organic wastes, especially animal wastes and organic effluents from the food processing industry. Diverse anaerobic bacteria participate in the degradation process and produce intermediate metabolites as well as endproducts such as methane, carbon dioxide, and ammonia.

The digested organic waste or slurry still contains organic matter, nutrients as well as microorganisms, and is superior over taw (undigested) slurry in terms of enhancing plant growth and improving the soil structure and nutrient status of soil. Methane is one of the major products, of the anaerobic digestion process as a fuel in the form of biogas. About twenty years ago, Rural Biogas Programs were developed and implemented and biogas digesters were built almost everywhere in the Chinese countryside; most of them are small and intended for household use, few of them being larger for use by the whole village.

By 1990, 4.7 million digesters had been built with a production of 1 billion cubic meter of biogas per annum. The farmers make use of the biogas for cooking, beating and lighting. Biogas use can save labor for cutting firewood, protect trees and groves, prevent the burning of crop residues thus increasing organic matter in soil, and generally improve farmers' living condition.

With time more advantages of anaerobic digestion were found.Digested slurry from biogas production is used comprehensively for fishing, cultivating mushrooms, and feeding livestock. It can increase crop output, augment farmers' income, decrease pollution, and hence attracts more farmers to use the technology. Anaerobic digested organic waste can be used to increase plant resistance to stress and diseases. For example, spraying supernatant of digested slurry on the leaf surfaces of orange trees can prevent freezing injury.

During the period of December 27-29 of 1991, temperatures in Jiangxi Province dropped to an usual low of— 7°C. Leaves of all Wenzhou tangerines in the area fell down because of the freezing damage and some of the trees died, except for one orchard run by a farmer who had sprinkled the surface of the leaves with the supernatant of digested slurry.

In next spring, the fruit bearing rate in that orchard reached 94.7%, while only 45% in the untreated trees. It has been shown that digested slurry can also increase cold-resistance and prevent decay of early rice, increasing the seedling survival rate by 8-15%. Compost has been successfully used to suppress certain soilborne diseases.

Digested slurry can decrease the incidence of several plant diseases, such as vegetable (cucumber, tomato, watermelon) and cotton wilt diseases caused by *Fusarium oxysporum,* cotton anthracnose caused by *Glomerella gossypii,* rice blast caused by *Pircularia oryzae,* rice sheath blight caused by *Rhizoctonia solani,* black rot of sweet potato caused by *Ceratocystis fimbriata,* rice seedling rot, and barley yellow mosaic diseases. Experiments in pot tests showed that a mixture of digested manure and soil decreases the incidence of watermelon fusarium wilt. High amounts of digested manure applied to soil had 86% control effect.

Digested solid residues applied to early rice seedling beds reduced rice bud rot by 52.5% and seedling rot by 24% comparing to the application of chemical fertilizers (control treatment). After transplanting, rice seedlings turned green and tillered 2-3 days earlier than the seedlings treated with chemical fertilizers. Soaking seed with anaerobic digested slurry increases germination power and germination percentage, and promotes seedling growth. The application of supernatant of digested slurry was found to be effective in controlling sucking pests, such as aphids, spider mites and planthoppers. The supernatant of digested slurry mixed with 450 g of Rogor emulsifier (40%) effectively controlled wheat aphids, as did 1500 a of Rogor emulsifier with water.

Applying digested slurry can increase the quality of food crops, for instance by increasing the sugar content of fruits and vegetables, and improving appearance and taste. Little is known about the mechanisms underlying the various effects of anaerobic digested slurry. Besides the metabolites mentioned above, digested slurry contains plant hormones such as indole acetic acid (1AA), cytokinin, and gibberellin.

Zhu and Wei studied the mechanism of enhancing cold resistance of rice seedling and showed that digested slurry can decrease ion oxosmosis of cells, increase free proline content of seedling, moderate the decrease of ascorbic acid and peroxidase, and increase the relative content of unsaturated fatty acids, factors and processes known to be related to increasing cold resistance. Inhibition tests in petri dish showed that digested slurry strongly inhibits the growth of the plant pathogens *Heblminthosporium, sativum* and *Rhizoctonia solani,* rid moderately inhibits the growth of the pathogens *H.turcicum H. maydis, H sigmoideum* and *Colletotrichum gossypii*. Shen found that ethylene and methane produced from digested slurry in microenvironments inhibited the activity of *Polymyza gracimimanis* responsible for the spread of yellow mosaic virus. Li et al suggested that the control of sweet potato soft rot in the storage period depends on the presence of the ammonia ion of digested slurry. Ma and Guo compared the physiological similarity of digested slurry and fulvic acid, a low aggregated humic acid. Both have the function of promoting plant growth, stimulating root system development, increasing cereal crop tiller, controlling rice sheath blight, aphids and spider mites, as well as promoting pig growth.

5 Bioindicators in Agroecosystem

Ecofarming is a term that will be used to indicate a kind of sustainable agriculture increasingly discussed and practised worldwide. Its main philosophy is to protect and stimulate soil life in order to increase or, at least, maintain soil fertility. Ecofarming derives from biology and ecology, especially the principle of the regulation of the abundance, distribution and activity of species in space and time. It is thus more comprehensive and rigid than the concept of sustainable agriculture which consists, as defined by Lchman *et al.* of agricultural processes that do not exhaust any irreplaceable resources which are essential to agriculture.

With regard to protozoa, investigations were performed mainly by Foissner in Austria and Zwart in the Netherlands. Both studied numerous biotic and abiotic parameters in several ecofarming variants in altogether 14 paired sites. Foissner studied testate amoebae and ciliates, Zwart heterotrophic flagellates and naked amoebae. Although different methods and indicator groups were used, the results match nicely and can be summarized as follows:

1. Many of the soil zoological parameters under investigation did not differ statistically in ecofarmed and conventionally fanned fields and grasslands;
2. There were no striking differences in species

composition and dominance structure of the ciliates and testate amoebae;

3. All differences that could be guaranteed with an error probability of α = 10% less invariably showed higher biological activity in the ecofarmed sites. The soil physical and chemical investigations which accompanied the zoological studies at some sites revealed that larger biological activity was correlated with a higher humus content and less soil compaction. The organic matter content was significantly higher in the ecofarmed plots, whereas soil compaction was more pronounced under conventional cultivation;
4. Conventional farming had a more detrimental effect on soil fauna in semiarid regions without animal husbandry than in Atlantic regions with mixed farming;
5. Agroccosystems are more sensitive to conventional farming than grasslands.

What is the meaning of the rather inconspicuous differences found? First, they show that conventional agriculture is often not as harmful to soil life as widely assumed. We have to investigate how these differences are related to crop yield and economic performance. Unfortunately, few studies are available relating crop yield and soil life.

The most comprehensive investigations were performed by Pfiffner's group in Switzerland. They compared three production systems in a 7-year crop rotation. Soil life was distinctly richer in the ecofarmed variants but crop yield and yield stability were decreased by about 20% as compared with the conventionally managed system.

However, yield loss was compensated by a 15% to 30% low total energy input per crop unit in the ecofarmed variants. The results of the Swiss group correlate well an earlier study by Lockeretz *et al.* who compared organic and conventional farms in the USA. They found that the slightly lower gross production per unit of cropland on the organic farms was largely offset by comparable reductions in operating expenses so that crop production was about equally profitable on the two types of farm, except in a year that had extremely favourable weather.

The exact comparison appears to depend on growing conditions, with the organic farms doing relatively better under the abnormally poor conditions of the mid-1970s, but relatively poorer when conditions improved in 1978. Except for wheat, a minor crop in the Corn Belt for which organic farms had much poorer yields, yields of most organically raised crops generally ranged from about the same to about 10% lower than on the conventional farms.

Along the same lines are very recent results by Drinkwater *et al.* and van Bruggen, who compared 20 conventional and organic tomato agroecosystenis in California. Both production systems could not be distinguished based on agronomic criteria such as fruit yield and arthropod pest damage levels.

However, differences were demonstrated in many soil, plant, disease, and diversity indicators suggesting that the ecological processes determining yields and pest levels in these two management systems are distinct. In particular, nitrogen mineralization potential and microbial and parasitoid abundance and diversity were higher in the organic farms. Differences between the agroecosystems were sufficiently robust to be distinguished from environmental variation and suggest that biological processes compensated for reductions in the use of synthetic fertilizers and pesticides. There are, however, also less favourable data available. Favretto *et al.* who compared mulched and tilled vineyard plots in Italy, found that green mulching significantly increased most nutrients and mesofaunal biomasses, but decreased yield by an average of 80% mainly due to water competition.

Along the same lines are the results by Foissner *et al.* who compared yield, fodder quality, and soil fauna in conventionally and ecologically farmed grassland plots in Austria. No significant differences were found for any of the parameters investigated, but labor input was higher for the ecofarmed variants than for the conventional plots. Obviously, results are at variance and strongly depend on investigation and farming methods, region, climate, and crops. Seen on a very rough scale, ecofarming appears as a promising practice, especially for sustaining biodiversity. As Lee and Pankburst concluded, at the present state of knowledge it seems wise to prefer management technologies that conserve the biodiversity of communities because these may provide the greatest benefits for the long term sustainability of the soil resource.

Literature on the effects of biocides on soil protozoa is comparatively extensive, but well-designed field studies are rare. Most of the commonly used biocides, e.g. atrazine, thiram, lindane, have been tested in at least one study. However the testate amoebae have been largely ignored, regardless of their high ecological significance and the reliable methods available for their abundance estimation. This is particularly frustrating in the light of recent studies which demonstrate more and more that a significant portion of the pesticides and their metabolites is bound to the humus fraction, a main source of food for testate *amoebae*. Furthermore, most studies did not consider effects at species level and used culture methods for abundance estimation, which poses many problems. Future research should thus concentrate on field studies, particularly testate amoebae, the most important indicator group.

1. The general pattern of reaction of soil protozoa to biocide stress is the same as that of other organisms;
2. Many protozoan species seem to be just as sensitive to pesticides as other more commonly used test organisms;
3. Insecticides are usually more toxic than herbicides.
4. Insecticides disturb soil protozoa critically, i.e. populations often do not fully recover within 60 days. Based on the recent literature, reviewed below, and some older data, mentioned in Foissner, these conclusions may be extended:
5. Fungicides have rather varied effects but most of them very likely do not disturb soil protozoa critically.

The following paragraphs discuss literature published in the years 1993 and 1994 as well as some papers not contained in my previous reviews. Chao and Alexander investigated the effects of the fungicides cycloheximide and thiram on protozoa in laboratory microcosms. *Rhizobium phaseoli* was more distinctly reduced by protozoa in the absence of the biocides, suggesting that protozoa were inhibited although they reached and maintained a rather high level.

Examination of the cycloheximide-treated soil revealed that the diversity of protozoa had declined abruptly, and only one morphological type was detected, a ciliate. However this could have been caused by methodological problems because culture methods are often and for inexplicable reasons.very selective. Griffiths studied the effects of a combined dose of cycloheximide and Triton X-100 on soil protozoa in laboratory microcosms. Both protozoan populations and nitrification were significantly reduced. Todorov and Golemansky studied the effects of three fungicides on three common, large freshwater and soil protists, viz. *Blepharisma japonicunt, Amoeba proteus,* and the testate amoeba *Arcella vulgaris* in liquid patch cultures. All species died within a few hours in 1% solutions of the biocides. Lavendotricin and Fuzamicin were more toxic than Fundasol, which even stimulated growth of *Arcella* at low concentrations.

The results of Todorov and Golemansky were not confirmed in a field trial performed by Miteva. She found that Fundasol was slightly more toxic than Lavendotricin. However, data were obtained with a selective culture method not distinguishing between active and cystic protozoa. Few differences could be guaranted statistically and numbers varied inconsistently, i.e. higher abundances sometimes occurred in the treated plots.

Differences did not become more pronounced when fungicides were combined with an insecticide, Agria-1060. No differences were found in species richness and composition of the, naked amoebae. Ekelund *et al.* studied the effects of two insecticides and a fungicide on total soil protozoa in soil microcosms and on three selected species in liquid patch cultures.

The biocides had a negative impact on the colonization of the sterilized soil microcosms by a natural population of soil protozoa. Fenpropimorph had an effect at the lowest concentration applied, i.e. 25 mg 11. However, treatment effects became insignificant for all biocides tested after 20 days. In single species cultures, *Cercomonas* sp. was the oranism most sensitive to fenpropimorph and pinimicarb, while Colpoda sp. was the most sensitive to dimethoate. The growth of Colpoda was apparently stimulated by very low concentrations of fenpropimorph or pirimicarb. *Acanthamoeba* was in all cases less sensitive than the other two organism. Ingham *et al.* studied the response of protozoa to five biocides, viz. streptomycin captan PCNB, carbofuran and cygon in a semiarid grassland soil in the USA. The substances were applied once in situ to soil in cylinders containing predominantly blue grama grass. Abundances of active and cystic protozoa were estimated with the culture method of Singh monthly between April and October.

In streptomycin treatments, cystic flagellates were greater in July and August. while totals were reduced in May. Total amoebal numbers in the soil depth of 05 em were initially increased in April but then recovered. In the 510 em depth of the streptomycin treatment, fluctuations in amoebal populations and the controls were asynchronic. In captan treatments total flagellate numbers were initially depressed but returned to control levels.

Cystic flagellate numbers were reduced significantly in June and August so that more of the flagellates were active. Total amoebal numbers were reduced until June in the 05 em depth, but then regained control levels. Only an initial reduction of amoebal numbers in the lower depth and no effect on cystic amoebae was observed. Ciliate numbers were not affected by captan.

PCNB initially increased (0-5 cm) or reduced (5-10 cm) all protozoa, followed by a return to control levels. Cystic flagellate numbers were greater in June and August so that fewer active flagellates were present on these two dates than in controls. Total amoebae in the 5-10 cm depth tended to fluctuate inversely to the control. Cystic amoebal numbers tended to be lower, although not always significantly, indicating that more amoebae were active in this treatment. Ciliate numbers also tended to fluctuate close to control numbers. Carbofuran did not affect flagellate numbers, either total or cystic, increased total amoebal numbers in April in the 0-5 cm depth but decreased them in the 5-10 cm depth, then fluctuated inversely to the control, generally ending lower than the control. In the 0-5 cm, total ciliate numbers were divergent frorn controls on two dates only, tenfold lower, 32 instead of 560, in May, and almost 100 fold higher, 2800 as opposed to 32 in controls, in August.

Flagellates, both total and cystic, were not affected by cygon. In both soil depths, total amoebal numbers fluctuated inversely to the controls. Cystic amoebal numbers were similar to controls, indicating that in May and June the amoebae were probably all encysted. In the lower soil depth, cygon generally reduced total amoebal numbers, so that the population was probably all encysted here also. Ciliates were generally not affected by cygon. Ingham *et al.* applied streptomycin, a bactericide, and captan a fungicide, to a non-tillage agroecosystem in Georgia, USA.

No significant treatment effects were noted for flagellates, naked amoebae, and ciliates, either in the litter layer or in the 0-5 cm, soil zone. However, the plots were analyzed only on days 7 and 14 after the biocide application and protozoan abundances were estimated with a simplified, selective culture method whose statistical soundness is doubtful Similar methodological weaknesses are inherent in another study, investigating the population responses of target and nontarget forest soil organisms to a set of biocides, viz. oxytetracycline-penicilline, captan, fumagillin and dimethoate-carbofuran.

The substances were administered to soil mesocosms at manufacturer recommended rates and treatment effects were analyzed only once. It after biocide application. All biocides reduced protozoa, especially fumagillin, and affected populations of nontarget organisms. MiteVa investiated the effects of the herbicide dual in strawberry plantations. The biocide was applied as recommended by the manufacturer and treatment effects were analysed fives times between Days 13 and 135. Protozoan abundances were estimated with a modified Singh method. Naked amoebae and ciliates, but not flagellates, were slightly decreased by the biocide.

No significant differences could be found in total protozoan numbers and species composition of naked amoebae. However, there was a constant trend in that numbers of amoebae, flagellates, and ciliates were slightly decreased in the treated plots. Thus, taking the five sample occasions as independent variables, a significant difference ($P < 0.\ 1$, at least) in total protozoan numbers would result. Parker *et al.* studied the effects of oxamyl and chlordane on litter and soil protozoa in laboratory microcosms. Protozoan abundances were estimated with Singh's culture method 5 and 30 days after administering the substances. Oxamyl caused a decrease of soil protozoa from 24.300X 104 g 1 soil to 5.440 X 104 g 1 soil between Days 5 and 30, but had no effect on litter protozoa. Chlordane affected neither litter nor soil protozoa.

Wanner and Wanner *et al.* performed many experiments on the influence *of* lime, a mineral fertilizer and two insecticides on a common soil and freshwater testate amoebae, Cyclopyxis kahli in liquid patch cultures, Insecticides were administered in two concentrations. Spruzit extended the lag phase significantly ($P < 005$), despite the fact that active pseudopodia emerged frorn C. *kuldi* immediately after addition of the biocide. Lower doses of spruzit and both concentrations of ripcord did not cause marked effects, Likewise, lime, fertilizer or lime and fertilizer did not improve culture growth, but both insecticides increased generation time significantly. Pesticides applied at 17°C caused larger shells of C. *kahli* as compared with the control.

6 PREDATORS IN AGROECOSYSTEM

Natural plant and animal communities, soil and water regimes have been altered as well in their structure and function as iii their dynamics. From an ecosystem point of view, agroecosystems resemble unmanaged systems by being solar powered, but differ because they have to rely on auxiliary energy to enhance productivity. Control of the system is external and goal-oriented. Under natural conditions, a succession would start immediately after a disturbance, such as deforestation by a catastrophic event, which would soon become a species rich ruderal community and within centuries a forest. When man clear-fells an area for planting, he does not allow natural development, but restricts duration of succession and rigidly controls the pioneer vegetation by planting only one species, which is the crop. With each crop, a controlled short term secondary succession is started. Early successional stages are nutrient-rich, with a high risk of nutrient loss when incorporation of nutrients into biomass by soil organisms is limited. Above and below ground species richness is low in the early stages of succession when a few specialized pioneer colonizers with high dispersal and reproductive capacities have the opportunity to build up large populations.

Diversity soon increases and may, in the intermediate stage, reach even higher levels than in the final successional stage. These general observations also apply to the small soil animals, the soil mesofauna. The importance of a rich soil mesofauna for ecosystem functions has been documented in many publications on faunal-microfloral interactions and nutrient cycling. Not only food web and nutrient cycling, but also the structural aspects of soil are positively influenced by these small animals.

The feeding interactions of arthropods in soil are reviewed by Werner and Dindal. Apart from pests, the effect of soil microarthropods on plant productivity is mainly indirect. Although the predators among them use only a very small amount of the system's energy, they may influence primary production considerably by controlling potential pests and by taking part in the rate regulation of nutrient cycles driven by the decomposer chain.

Agroecosystems are frequently disturbed and succession is started anew. Consequently, for understanding agroecosystems, ecological studies from early successional stages are of great value. When a field is prepared for a new crop, ecosystem development is controlled to a large extent by abiotic external factors. With the establishment of populations interactions within the system become more and more important. Nematodes, for example, have been found to be fast colonizers and their predators soon follow and interact with their prey populations.

General predator-prey models support the observation that control by predation is very efficient, particularly at low prey levels, even for prey with very high reproductive potential. Mites from the group Gamasina are one important group of predators involved in these internal control mechanisms. The above-ground structure of the vegetation layer greatly influences soil mesofauna abundance and diversity. This has been proven in field and mesocosm experiments. The effects are indirect by providing a favorable microclimate for the development of soil fauna, by influencing their food, e.g. microflora and by increasing vertical and horizontal patchiness and habitat diversity. The importance of diversity for decomposition has been shown by Verhoeff and Brussaard, who concluded that different assemblages of species may be involved in the same process.

Lawton discussed several hypotheses on the implications of species richness and ecosystem function (necessity of minimal diversity beyond that redundancy; all species contribute to ecosystem performance; ecosystem function changes unpredictably when diversity and/or species composition changes). The impact of agricultural practice on soil biota and their functions very much depends on the cultivation method chosen. Ploughing buries the organic material into deeper soil, where under well aerated conditions fast metabolizing bacteria may thrive., whereas no-tillage favors the development of fungi. Thus, the decomposer food chain in a conventional tillage system is bacteria-based, with a high risk of loss of nutrients to leaching. In a no-till system it is fungi-based and more nutrient conserving. For soil biota, ploughing is an elementary catastrophy and causes loss of diversity and population density.

In contrast to conventional cultivation with its rigid external control mechanisms, integrated, sustainable or 'alterative' agriculture recognizes the advantages of natural processes and tries to protect soil life and its diversity. Availability of organic material and a favorable microclimate may be provided by appropriate weed control or mulching. In many cultures, e.g. vineyards, management of habitat diversity in terms of a two or multilayered vegetation structure has been successful in increasing microarthropod diversity.

Concepts of integration provide the tool for lowering energy subsidies and increasing ecological diversity. Tillage techniques are among the most important management tools for achieving sustainable productivity. An ecological agriculture is only possible in a sound economical framework. The economic benefits of an integrated agriculture are well documented by Pimentel. The assessment of the importance of diversity and the presentation of solutions to the very specific problems of individual farms still requires tremendous research effort. Even mere documentation of diversity may cause difficulties, since it is based on counts of numbers and identification of systematic groups (taxa) of organisms.

A specialist will be able to estimate diversity at species level, and will find high values compared with an other researcher who counts on the level of an order, such as mites or spiders. Therefore, the determination of organisms (taxonomy) is a prerequisite for studying diversity. The value of lumping together species in functional groups or guilds is very much questioned by the work of Walter and Ikonen. With overwhelming species richness and high abundances within the range of hundreds of thousands of individuals per m, mites and Collembola are the most important microarthropod groups. Among the mites, predatory Gamasina are a group which has an important controlling function and which may be used for indication of soil quality and anthropogenic impact.

Mites (Acari)

Mites can be found in virtually all habitats, front the sea to the glacial zone, and in the soil, even in ground water. Almost any food source is used, and a great number of species are ecto or endoparasitic. This remarkable diversity is reflected in numerous morphological and biological specializations: 40000 mite species have been described tip till now: many more are expected in the future. By their high diversity and often great numbers, they are deeply involved in many ecological interactions. In the last decades, they have attracted increasing interest in the context of decomposition, pest problems and their control, bioindication and human health. With body length of rarely more than 2mm, the mites are the dwarfs among the spider-like arthropods. The small size not only has secluded them for centuries from close inspection, but also involves many specific morphological peculiarities. Like the other arthropods, they possess a chitinous exoskeleton, which may be soft or sclerotized to various degrees.

Sclerotized parts of the limbs and plates of the body are connected by soft membrane. Hairs of various shape and number originate from the cuticle. They are arranged according to specific patterns. Body division, number and shape of plates and hairs, and type of setal arrangement are important identification features.

The order Acari is divided into the two suborders Acariformes and Parasitiformes. The Acariformes comprise three large systematic groups, the Acaridida (Astigmata), Actinedida (Prostigmata) and Oribatida (Cryptostigmata). The -stigmata and -ida terminologies are only roughly identical. The Parasitiformes include three smaller systematic groups (Opilioacarida, Holothyrida, Ixodida (ticks) and one larger group, the Mesostigmata or Gamasida. Apart from some external or internal parasites of animals, the Parasitiformes are for their majority predators. Mesostigmata possess one pair of stigmata connected with an anteriorly directed groove, the peritreme. Four free-living developmental stages are found: larva; protonymph, deutonymph; and adult.

The group is divided into two systematic branches, one of which combine, the Sejina with the Uropodina and

Gamasina. Sejina mainly live in tropical forest litter. Gamasina and Uropodina include many important species of agricultural soils all over the world.

Uropodina

Uropodina are found in large numbers in the range of 10000 m^{-2} in highly organic materials, such as compost, manure, older sewage sludge, decaying stumps or deciduous forest litter, where they prey mainly on Nematodes, insect larvae, unfit or even decaying earthworms. In less organic substrates, such as meadow or agricultural soil, Uropodina abundances are low. Species number per site is generally in the range of ten or less. Distribution of Uropodina species is quite specific to habitat quality. Increase of pore volume and soil organic matter is reflected in rich uropodine communities. The preferred habitats of Uropodina (decaying organic matter with many Nematodes) are mostly ephemeral, and the mites have to migrate when their habitat has been decomposed. Only deutonymphs migrate. They exhibit a specific behavior before they adhere to insects which bring them to a 'fresh' habitat. An adhesive stalk or tail is produced from the anus to attach to the carrier, a feature which is included in the scientific name of the group (uro: tail; pous: foot). Raut and Panigrahi suggest the use of Uropodina in biological control,: the mites have been observed in attacking slugs and killing them by making holes in the dorsal surface of their body.

Gamasina

The majority of the Gamasina are mobile predators. Gamasina have no eyes, but there is an overall sensitivity to light, which they avoid. They detect their prey by chemical or tactile stimuli. Volatile kairomones of spider mites, for example, are detected by Phytoseiidae. Most of the sensilla are located on the pedipalps and on the first pair of legs. The pedipalps probe the surface of the ground, apparently for orientation and chemical stimuli. The first pair of legs serves as a feeler and is used to fix the prey by pressing it to the ground for piercing or cutting the skin with the chelicerae. Prey fragments are not swallowed, the predators digest their food preorally by producing digesting liquor, which dissolves the prey and is sucked up. Eggs may be produced sexually or parthenogenetically. A six legged larva hatches, which after moulting to protonymph and deutonymph develops to the adult stage with sexual organs. Life history data vary considerably, e.g. development times from egg to adult have been reported from 2 (36°C) to almost 200 (25°C) days, longevity from 5 (20°C) to more than 300 days (15°C), for different species from various sources cited in Ruf. Information on temperature is important, since life history data are very much influenced by microclimate.

In open landscapes, Gamasina occur in the range of 10000 m^{-2} in undisturbed meadows (0-15 cm depth) and even more can be found in disturbed soils or in early successional stages such as dunes or ruderal sites. Species numbers are in the range of 15 for undisturbed open grassland sites; ruderal sites may yield more than 25 species and forests more than 30. Gamasina are typically predators with the exception of some genera from the superfamily Phytoseioidea, which live in the herbal layer and feed on pollen.

Epedaphic Gamasina are large (ca. 0.8-1.5 mm), long-legged species, well sclerotized and yellow-brown in color. They hunt on the soil surface in the litter layer for other small arthropods, particularly Collembola. To a large extent, they belong to the families Ascidae, Parasitidae, and Veigaiaidae. Others feed on eggs of insects and their first instar larvae, preferably in habitats such as compost, animal droppings or manure (many

Macrochelidae). Euedaphic species live in the soil pores. They are small and slender (ca. 0.25-0.5 mm), whitish in color and with rather short legs. The typical soil dwellers from the family Rhodacaridae have a divided dorsal shield, the anterior of which may be partitioned again to accomplish maximal flexibility in the narrow and crooked pores. Soil dwellers probably do not have much success in hunting in the pores by running. Rather, their hunting ability may depend on availability, i.e. on the density of prey in the microhabitat.

Nematodes and other small wormlike food generally abound in the soil and are their preferred prey. These animals are seized with the chelicerae when they stick their bodies into the air filled lumen of the pores. When there is mass occurrence of soil microarthropods, such as 'root mites' (Rhizoglyphus, Astigmata) or Onychiuridae (Collembola), these may also be eaten by Gamasina. Feeding preferences are widespread and even specialization occurs, as in *Alliphis siculus*, which exclusively feeds on Nematodes. The form of chelicerae has been associated with feeding preferences by Karg. This view is supported by weak correlations of morphometrical parameters of chelicerae and food preference by Buryn and Brandl and opposed by Walter *et al.*

In a rural landscape with hedges in Southern Germany, Buryn and Hartmann found the highest diversity of soil Gamasina at the edge between the cultivated meadow and hedge, followed by hedge and meadow.

Nematophagous species were dominant in the meadow and microarthropod feeders in the hedge. The hedge community included typical forest species. Also for other habitats or successional stages, specific species associations have been found, but autecological studies are too rare to find causalities for the existence of these communities. For recently disturbed sites with ruderal vegetation and low humus content *Arctoseius cetratus*, species of the family Rhodacaridae, particularly *Rhodacarus silesiacus* and some surface species, like *Parasitus eta* or *Veigaia nemorensis* are characteristic. If such soils contain moderate amounts of organic material. *Alliphis siculus* is found regularly and Macrochelidae are typical colonizers of fecal droppings. Recolonization of disturbed sites very much occurs by wind drift and insect transport.

As described for the Uropodina, phoretic distribution is also found particularly in deutonymphs of Gamasina species which colonize ephemeral habitats., such as dung; special morphological features of migrating stages do not exist. Experiments showed that diaspores in topsoils even persist under treatment by heavy machinery, transport and desiccation. They can develop under favorable microclimatic conditions. Recolonization also may be fostered by hedges with their species rich edges and by islands with preserved sod in an otherwise heavily worked in agricultural landscape.

Like other soil mesofauna, population development of Gamasina is very much influenced by microclimate, which depends on the structure of the herbacious plants, shrubs or trees and on the litter layer. Removal of litter reduced species number in a forest from 21 to eight species.

Experimental increase of soil moisture and reduction of daily temperature amplitude by cages covered with a 1 mm mesh raised abundances by a factor of five, which was mainly due to *Arctoseius cetratus*. This experiment was conducted on an early successional site and no major changes of species spectrum were observed.

Gamasina do not change soil structure or plant productivity directly. However, as predators, they influence population growth of other organisms and through this have an indirect effect on overall ecosystem performance. The multiple interactions with their prey are pretty diverse, not only with respect to their food, but also with respect to the soil profile and the patchiness of habitat.

Nematodes play a central role as prey for Gamasina in agroecosystems. Nematode pests may be controlled very efficiently by these mite predators. Experiments by Bund document Nematode reductions of almost 70% due to predation by *Hypoaspis aculeifer*. Much evidence for the effectiveness of soil Gamasina in holding in cheek potential pests is reported by Karg. Conventional agricultural systems are characterized by frequent and repeated disturbance, often within the mite's life cycle, and by 'clean' soil surface, particularly between harvest and cultivation. The depth of disturbance is considerable when a plough is used. As early as in the first half of this century. Franz published drastically lower abundances of soil fauna in fields, compared with uncultivated landscapes and on the consequences for soil fertility.

The same trend has been observed by various authors for Gamasina. Compared with forests, Gamasina species diversity in conventional fanned fields, may be reduced by more than 75% to species numbers in the range of five. Besides the mechanical disturbance, this drastic impoverishment can be related to the lack of a litter layer, which does not allow the development of a diverse prey spectrum to feed the predators.

As in 'natural' sites, specific communities are associated with the peculiar environmental conditions of agricultural fields. Survival of ploughing or other structural 'catastrophies' (from the point of view of a soil organism) or fast recolonization with high tolerance towards microclimatic fluctuations in bare or sparsely grown fields, are some of the selecting requirements.

Small euedaphic species with rather high reproductive capacity are able to cope with these conditions, e.g. Rhodacarellus silesiacus, Arctoseius cetratus. These species are found not only in fields but also in early succession of ruderal sites, in sandy soils low in organic material. Fertilization with manure or dung allows colonization by phoretic surface species, e.g., fro the families Macrochelidae and Parasitidae. Lack of predation pressure by Gamasina may result in the development of pests of otherwise harmless organisms. In a 'clean' monoculture, there is nothing else to cat for these populations than the crop. Considerable damage to the farm's economy is inevitable, either as loss of yield or as cost for pest control, or both.

In a 3-year study, EI Titi found that ploughing has the most negative effect on soil Gamasina abundance and species diversity, compared with rotary and broadshare cultivator. Besides reduced mechanical disturbance of the soil, no-tillage cultivation does protect the organic layer of the field. By this management technique, abundance and Gamasina diversity can be raised by roughly 50%. However, no-till cultivation has to rely on herbicides to control weeds.

Many herbicides persist and may cause reductions of population levels, as was observed in relation to an old field site by Moore *et al.*, 1984. It must be noted that old field sites may be very species rich with high abundances. The other effect is related to the potential pest, for which alternative food may be provided. It is not only tillage technique that has an effect on Gamasina diversity, but also soil compaction by machinery. Heisler performed experiments with increasing compaction.

Strong compaction in the wheeltrack reduced abundance of Gamasina by 40% and species number by 30%. Epedaphic Pergamasus species, Alliphis siculus and euedaphic Rhodacarellus silesiacus are affected most. Arctoseius cetratus is resistant to compaction and becomes by far the most dominant species. Such a high numerical concentration to few or even one species reduces the efficiency of the Gamasina community as controlling predators. Mineral fertilization, particularly in combination with manure and moderate use of sludge with low heavy metal content has a positive indirect effect on Gamasina abundance by stimulating population development of prey organisms; diversity is not significantly affected.

Pesticide application triggers complex reactions of the soil zoocenosis. Gamasina as predators integrate effects, which may be indicated as changes of abundance and species spectrum. The mites may suffer from direct toxic effects or indirectly from changes in the prey populations. In an experiment, the succession of Gamasina in the soil of two synchronous experimental plots (undisturbed succession, sown grass with regular mowing) was observed with and without initial Aldicarb application. On the treated sites, species numbers did not recover from the impact within one year and community structure was strongly concentrated to only one or two species.Of the controls, dominances were not so pronounced. A variety of species specific reactions were observed for Gamasina: total extinction; late recoverage; effect depending on type of vegetation; stimulated develop merit; and indifference.

Similar to the results of EI Titi, Rhodacaridae were most affected, which is related to the water solubility of Aldicarb. Stimulation of surface dwelling species may be due to an altered competitive situation and to increased hunting success because of slow, only partially poisoned prey. The experiment also confirmed indirect effects of reduced predation pressure on prey: since the abundance of Rhodacaridae was strongly reduced after the pesticide application, a collembolan prey population *(Mesaphorura krausbaueri)* developed unchecked. Another indirect effect may be possible causing the strong decrease of the nematophagous *Alliphis siculus:* Aldicarb is very nematicidal and *Alliphis siculus* suffers from lack of food.

From the feeding biology and from their interrelations in the ecosystem, main areas of pest control with Gamasina can be identified: Nematodes; Collembola; insects; and pests of the shrub layer. Since the early 1960s, Gamasina have been seen as an efficient biological control of Nematoda. Above ground, Gamasina of the family Phytoseiidae are important antagonists, particularly of spider mites. In this field much experience has been acquired, e.g, on the right timing of the release of the predator.

Macrochelidae feed on eggs and larvae of insects. In poultry houses, housefly populations have to be controlled. Macrochelidae have been shown to effectively control these insects, when some prerequisites are observed, such as alternate prey to overcome lack of main prey, and leftovers of manure for Macrochelidae to breed. The few examples show that successful biological control must be based on good knowledge of the biology and ecology of the species involved.

In a monograph on Gamasina, Karg gives an introduction to the group and a key to more than 1000 species from Central Europe. Either intact or mixed soil samples are collected from the field. Care should be taken to separate minor habitats, such as soil horizons and patches of organic matter, distance to crop plant, within and between planting rows, etc. Gamasina are expelled from the soil by dynamic extraction, which takes advantage of the mobility of the animals. The principle of the extraction method is simple. In the laboratory, the soil conditions should be made comfortable for the animals, so that they are active (humid and cool).

For mixed samples the soil should be handled gently to avoid crushing the mites. The soil is placed into a sieve over a funnel and dried carefully over a couple of days by applying increasing heat, e.g. with a dimmable electric bulb.

By this method, a moisture and heat gradient of ca. 10 K should be achieved between the surface and the bottom of the sample. Towards the end of extraction, this gradient is reduced and final temperatures of ca. 400°C at the bottom of the soil sample should be reached to make mites leave this zone. The soil animals fall via the funnel into a glass with preservation fluid (70% ethanol) and can be counted with a stereoscope. Cleared specimens are needed for identification with a microscope, and permanent mounts are very useful.

The distribution of soil animals is aggregated and high variation of abundance estimates in the range of 60% of the mean are to be expected (P= 90%). These values have been derived torn the analyses of extended data sets for a widely used sample size of $n = 10$. Assessment of species spectrum is more reliable.

Gamasina are caught alive either with dynamic extraction or in the field with a fine brush. The mites are collected into humid vessels, e.g. with a moistened bottom of a mixture of plaster of Paris and charcoal (1:1). Gamasina can be fed with nematodes obtained from horse manure, with relatives of cheese mites (Acaridae) or with potworms.

7 Spider Biodiversity

Spider communities were studied in different agroecosystems throughout the world. Few, however, were carried out in Europe, and in particular the arborescent layer was neglected. Not a single study was done in France. Our work was carried out in different arborescent environments such as pine, oak and apple trees, but we limit our discussion here to the experiments concerning apple orchards, which are of greater economic importance.

Every year, this agroecosystem suffers severe damage due to arthropod pests, which can be classified into five groups. The purpose of this study is to show that although Araneae can be considered as polyphagous predators if the group is treated as a whole, they are to be looked at as specialised predators if their hunting strategies and location in the vegetation are taken into account. As a consequence, not all species are useful against a particular pest, but preserving their diversity could be important for controlling different pests.

Spiders were sampled in the trees of two chemically untreated apple orchards, about 30 years old in the western part of France mainly by means of two method:

— beating branches during spring and summer and collecting Arancae using an entomological umbrella and a poorer. This method was used during the day;

— putting cardboard traps on tree trunks before the cold season. These traps, 10 em wide, were wrapped around the trunks of trees at a height of 50 em and covered by a metallic protection to prevent birds destroying them. The traps were placed in the orchards at the end of July and collected at the beginning of the cold season. When collecting, a bite piece of cotton was placed all around the base of' tree to prevent spiders from escaping. The individuals that fell down when taking off the traps were caught using a pooter, and the traps were packed in plastic bags.

These two method were complemented by visual searching: just collecting spiders using a pooter. All spiders collected were preserved in alcohol for subsequent identification in the laboratory. About one thousand individuals were collected and 39 spices were identified belonging to 14 families. Of these, ten were only found on, or under, the bark of the trunk. Some of them usually live in the grass layer and move onto the trees in order to find a shelter to pass the cold season.

Unfortunately the collecting effort and the number of orchards sampled is not always specified, but it does show the generally high diversity of these predators. In spring and summer, four families accounted for more than 80% of the foliage *spider community.*

The Araneidae (orb-web spiders). Theridiidae (frame-web spiders), Philodromidae (ambush spielers) and Dictynidae (frame-web spiders). One family the Clubionidae (nocturnal wandering species), represented almost two-thirds of the community on the trunk of apple trees. Together with the families of Dictynidae and Philodromidae, these make up over, 85% of the community.

First it should be noted that, in some conditions, the branch-beating method may underestimate two families, the Salticidae and the Clubionidae. This depends on the moment of sampling in relation to their biology and activity. These Arancae build a silken retreat attached to leaves or branches in which they stay at night or even during the day, which might prevent them falling down into the entomological umbrella when the foliage is shaken.

When sampling during the day time, the Salticidae, a diurnal wandering species, might be slightly underestimated if the weather is very cloudy because some individuals will remains in their nest. The Clubionidae, a nocturnal wandering species are certainly underestimated under such conditions.

This explains at least part the low number of Clubionidae found in foliage during spring and summer. Notwithstanding these disadvantages, this sampling technique gives satisfactory quantitative results for the estimation of spider communities on apple trees. The differences between the foliage and trunk communities of spiders are explained primarily by, the structure of the habitat, as their localisation on one of these tree subhabitats is linked to the type of hunting strategy used by the predator.

Branch beating has shown that spiders are scarce in the upper part of the tree during the cold season. Indeed, before winter, the spiders move down the tree, trying to find a shelter under the bark, and their activity is extremely reduced for almost all of northern Europe.

It has been demonstrated in the Clubionidae that there is a seasonal migration from the foliage to the bottom of the trees before the cold period. When the weather becomes more favourable, usually at the beginning of spring, they move back into the foliage. The structure of the bark is important because it influences the presence of some of these Arancae on the tree, especially in large species that require a sufficient space under the bark to construct their winter resting nest .

Such migrations are also clear in ambush spiders, such as Philodromidae, which are well represented on trees, and in diurnal wandering spiders which are not numerous in such environments as they are species mainly of open and sunny areas. Some species pass this period as eggs (mainly Araneidae) and first juvenile instars (Theridiidae), explaining why they are so poorly represented in the trunk community

Most of the spiders live among the foliage from spring to autumn. Spiders using a web to capture their prey need to find a structure to support the construction of the trap. So, except during winter, the majority of them (except tubular-web species) colonise the foliage in different places according to the type of web (orb-web, frame web or sheet-web) and the size of the individual. The orb-web spiders particularly colonise the space between fairly large branches which are necessary to establish their more or less vertical sticky webs, situated at all levels in the foliage.

The frame-web spiders (e.g. *Lathys humilis*) colonise the extremities of small branches, building an irregular three-dimensional web that links adjacent small branches of a group to each other. In this group, a few small Dictynidae species are able to construct their trap on the trunk if the bark is partly covered with lichens.

The sheet-web of the Linyphiidae, found at the lower level of the trees' foliage, is composed of a horizontal part suspended by several vertical, non-sticky silken threads. So, during the favourable period of the year few spiders are located on the trunk of apple trees, and their numbers are lower the smoother the bark is.

However, the tubular-web species (e.g. *Segestria bauarica)* are never found in the foliage, but only on the trunk, because they need the bark structure of the trunk to build their web under. The silken tube is protected by the bark and only the entrance can be seen from the surface of the trunk.

The bottom of the trunk also provides a few suitable places for some small Linyphiidae to establish their web, and for some nocturnal wandering spiders, such as Gnaphosidae, which more often occur in the grass layer. A few Philodromidae can also be found on the trunk and large branches.

Although the localisation of the main species of spiders on apple trees during spring and summer depends on their hunting strategies the whole tree is colonised. The structure of the environment is very important, especially for the web-building species.

The biology and different hunting strategies of the Araneae have an effect on the species distribution in the environment and in consequence on the type of prey consumed. In apple orchards, Lepidoptera are captured by nocturnal wandering spiders, as well as by ambush predators and orb-web species.

The larvae are mainly consumed by the nocturnal wandering spiders, but also by ambush species as Philodromidae. The efficiency of each group needs to be considered in relation to their hunting strategy. Indeed, in Philodromidae the displacements are lower than in Clubionidae, hence their probability of encountering larvae is lower.

Butterflies will principally be captured by orb-web species which have a web solid and sticky enough to resist and trap the adult Lepidoptera when flying through the foliage in the evening. Ambush species like Philodromidae only catch mating adults. *Anthonomus pomorum* is only consumed by adult spiders. This pest species is trapped by frame-web spiders occupying the extremities of the branches with their web, where the females of this beetle come to lay their eggs in buds.

8 FARMING WITH BIODIVERSITY

Agricultural progress seems even more paradoxical when we consider biodiversity use in agriculture. Several thousand plant and animal species are known to have served as food in human history. The displacement of traditional crop varieties by newly-developed high-yielding varieties has been a major factor in this development. FAO estimates that in the next 10 years 90% of the existing genetic diversity within the main crops is at serious risk of disappearing. The vast majority of the literature dealing with biodiversity use in agriculture is characterized by a symptomatic approach, and centers on providing data that prove the continuous reduction of both inter and intra-species biodiversity used in food production, on its associated risks for the sustainability of our food supply, and on alternative farming techniques to halt the erosion of biodiversity use in agriculture.

A discussion of the driving forces that have generated the reduction of biodiversity use in agriculture is generally missing. Put another way, the detection of symptoms is directly coupled, to a prescription of remedies without any serious attempts to diagnose the disease. Perhaps this is because a thorough diagnosis would necessarily raise some politically unpleasant issues. The decreasing energy output/input ratio in agriculture means that the food security of humankind is increasingly dependent on rapidly dwindling oil reserves, and on an increasing environmental impact per unit of crop output. The dramatic reduction of biodiversity worldwide spells serious problem for the maintenance of ecological, equilibria at the planet level. That is, the path of technological in food production chosen right now by humankind is not sustainable. Not only are these unpopular issues to discuss, but it is becoming evident that most of the undesirable side-effects leading to unsustainability are difficult, if riot impossible, to incorporate in conventional sets of measures and indicators used to assess risks, costs and benefits of technological change. In spite of these difficulties, it is crucial to understand the driving forces that are shaping current trends of technological development in Otherwise, we seriously risk caveats in discussions of future agricultural policies. It is common knowledge that traditional techniques achieved high output/input energy ratios and used much more biodiversity than modern high-input agriculture.

Individual scientific disciplines that study changes in socioeconomic systems tend to focus on the effect of one factor at the time. Although this approach provides a high resolution for the factor in question, it does not shed much light on the bigger picture. In this respect the agricultural sciences are no exception.

Ecologists tend to focus on the on-site and off-site effects of agriculture on the environment, without considering agriculture as a livelihood. Economists are concerned with the economic return of the labour and capital resources invested in the farming process, and tend to more the ecological impacts. Agronomists occupy themselves basically with the physical performance of the production system. Rural sociologists and anthropologists are mainly interested in the effects of technological changes on the lives of individual farmers and the identity of rural communities. This specialization or *reductionism* of established scientific disciplines is at times necessary, but often creates confusion for decision-makers that have to decide about the future directions of technical development in agriculture.

All of the above-mentioned aspects of agriculture must be addressed, since each one is fundamental for the sustainability of agriculture. The problem lies in the fact that a detailed discussion of any one of these aspects becomes useless if not put into relation to the functioning of the system as a For this reason, it is of utmost importance to put the different disciplines in perspective within a common frame of analysis.

The existence of different perspectives for assessment of stistainability of a agricultural development has been pointed out by many workers. In spite of minor differences in definitions and emphasis among the different authors, there is substantial agreement that at least three fundamental perspectives should be considered when attempting to define sustainable agriculture:

- The ecological view: agricultural techniques must be environmentally sound;
- The economic view: agricultural techniques must be economically viable;
- The social view: agricultural techniques must be acceptable to farmers and society given culture, ethics and religion.

Assessing only agronomic performance or ecological compa-tibility of various agricultural techniques is not sufficient to evaluate their sustainability. We need a parallel check on economic viabilily and social acceptability. 1 believe that it is thus safe to conclude that we need to focus our efforts on the development of frameworks or models of analysis that establish bridges among the different perspectives. The farming system as a sub-system of society: constraints imposed by population density, income level and labour time allocation in society. In order to verify whether certain agricultural production techniques are convenient for a given society, we should first analyse how the organization and technological development of society influence the way in which food is produced in its agricultural sector. Two main characteristics of society that I will examine in this paper are:

1. Population density, or demographic pressure. One of the most obvious aspects of demographic pressure that directly influences the agricultural sector is the amount of arable land per capita available for agricultural production.
2. The level of technological development of a society, or socioeconomic pressure. This can be conveniently quantified by the following two indicators:

- The average per capita income or average per capita commercial energy consumption ill society. For instance, average per capita commercial energy consumption ranges from less than 1 GJ per year in developing countries to more than 300 GJ per year in the USA.
- The fraction of total human time available that is allocated to work in the primary sectors of the economy. This fraction ranges from 0,1 ill developing countries to 0.04 ill developed countries. The lower fraction of human time allocated to the primary economic sectors in developed countries is related to the longer education period, the progressive graying of the population, the lighter

workload of the labour force, and the expansion of the service sector (up to 60% of the labour force). Of the primary economic sectors, agriculture is one of the most affected in terms of contraction of human time allocation. A cross-section analysis over a sample of 60 countries for the years 1990-1991 shows that the fraction of the labour force engaged in agriculture is inversely correlated with the GDP per capita. Countries with a GDP per capita of more than US$10000 per year have less than 7% of the working force in agriculture, and the richest countries even as little as 2%.

Do population growth (increase in demographic pressure) and changes in energy consumption and time allocation pattern in the economic process (increase in socioeconomic pressure) have implications for the acceptability of techniques of food production in a developed society? In the affirmative, are there numerical indicators that can express the demographic and socioeconomic pressure that society exerts on the production techniques adopted at the farm level?

The demographic pressure or population density of society is simply defined as the ratio. It is well known that population density has been an important factor in shaping the development of agricultural production techniques. Apart from scientific analyses, common sense also suggests that a high demographic pressure in society will favour farming techniques that yield a high food production per unit area.

The effect of demographic pressure on agricultural production can be expressed by the indicator AP_{DP} (= agricultural productivity imposed by demographic pressure), which is defined as the productivity of land (yield of food energy per hectare) 2 that would be needed to obtain a situation of complete food self-sufficiency in society (matching the aggregate food requirement of society). Thus defined numerical assessments of AP_{DP} require knowledge of the aggregate requirement of food in society (considering the food system under analysis as closed), which is in turn determined by the population size of society, the food consumption pattern, and post-harvest losses; The amount of land available for food production, which depends on availability of arable land, characteristics of arable land, and alternative land-uses (dependent on population size and level of technical development).

The effect of socioeconomic pressure on agriculture can be expressed by the indicator AP_{SEP} agricultural productivity imposed by socioeconomic pressure), which is defined as the productivity of labour in agriculture (yield of food energy per hour of labour in the agricultural sector) that would be needed to obtain food self-sufficiency for society.

Considering society's food system as closed, AP_{SEP} is obtained by dividing the yearly aggregate requirement of food in society by the labour time available in a year for generating food supply. The labour time available in a year for generating the food supply depends on the size of the labour force, the unemployment rate, tile fraction of the labour force absorbed by the non-agricultural sectors, and the average work load. The US economy, with only 2% of its workers engaged in agriculture, has an APSEP of' 270 MJ of food energy per hour of agricultural labour. This value is well out of the range of possible productivity of fanners in developing countries (less than 10 W 1 and is not even reached by farmers in the European Union. The requirement for food in society is proportional to population size, and therefore to the total human time available to society, whereas internal food production is proportional to the labour time available in the agricultural sector. Hence, in order to achieve and maintain food self-sufficiency, any reduction in the fraction of total labour time allocated to the sector must be compensated by an increase in the amount of food produced per hour of labour in the agricultural sector. According to this rationale, a socioeconomic structure that allocates only a small fraction of its labour force to the agricultural sector will select farming techniques that result in a high food production per unit of labour in the agricultural sector.

The indicators AP_{DP} and AP_{SEP} represent the minimum density of biophysical throughput, per hectare and per hour, respectively, that would be needed on average in the agricultural sector to achieve a state of food self-sufficiency in a defined society. Hence, they are based on characteristics of the socioeconomic system as a whole, and their values can be calculated without any knowledge of what is going on at the level of the fanning system in the society. How, then, can these indicators be related to the actual technological changes occurring at the farm level? Are the values of the indicators AP_{DP} and AP_{SEP}, that characterize the pressure that society exerts on its fanning systems, effectively related to the techniques of food production realized at the farm level? To answer these questions it is useful to compare, for several types of societies, the values of AP_{DP} and AP_{SEP} with actual values achieved in agricultural production in these societies. For instance, the minimum productivity of land imposed by demographic characteristics (AP_{DP}) can he compared with the actual agricultural productivity per hectare (AP_{ha}) for the country as a whole (average), or at the regional level, or for individual farms. The actual productivity of land or yield of food energy per hectare achieved by a country, or farming system is obtained by dividing the aggregate production of food within the defined territory by the amount of land used in the production process, Similarly, expected minimum value of labour productivity in agriculture according to socioeconomic characteristics can be compared with the actual agricultural productivity (throughput of food energy) achieved per hour of labour in agriculture for the country as a whole (average) or for selected farming systems. Such a comparative analysis, at the national level has been carried out for a large sample of countries, to see whether the demographic and socioeconomic pressure are indeed related to the way in which food is actually produced in different socioeconomic contexts. The countries included in the sample are listed in Table 1 and cover a wide range of demographic pressure, socioeconomic pressure, and geographic location.

The analysis shows that even if the two indicators do not indicate mandatory threshold values, they should in any case be considered attractor values, toward which the agricultural system feuds to move in practice (Figs. 2 and 3) 3, If a country has a high demographic pressure it is also likely to have a large production of food energy per hectare in the agricultural sector.

In other words, current technological performance in agriculture in terms of yield per hectare is affected by the existing demographic pressure. Similarly a country experiences considerable socioeconomic pressure it is also likely to have a large production of food energy per hour of work delivered in the agricultural sector. Hence, technological performance in agriculture in terms of labour productivity is affected by the existing socioeconomic pressure. Note that in industrialized societies trade plays a significant role in stabilizing an equilibrium between the requirement and supply of food. Whenever AP_{DP} exceeds AP_{ha} at country level and/or AP_{SEP} exceeds AP_{ha} at country level, society somehow relies on import to offset existing differences between internal food requirements and the internal supply of food realized (both in terms of land and agricultural labour), Obviously, countries with net food export have a value of AP_{DP} that is smaller than their AP_{ha} arid/or a value of AP_{SEP} that is smaller than their $AP_{hour.}$ If the analysis is restricted to countries that had a GNP per capita of less than US$1000 per year in 1991, it becomes clear that in poor countries, where the food system is less open and hence risk aversion is of greater importance, food self-sufficiency in terms of both land and labour productivity in the agricultural sector is more strictly adhered to in practice than in industrialized countries where trade plays a major role.

Countries not achieving self-sufficiency must rely on imports or food aid to guarantee food security for their population. The analysis described above indicates the existence of a link between the variables AP_{DP} and AP_{SEP} on the one hand, that are defined at the hierarchical level of society as a whole, and the variables AP_{DP} and AP_{hour} on the other hand, that are defined and measured at the hierarchical level of farming systems. This link suggests that it is possible to describe technical changes in food production on both levels on the same geometric plane. This plane is defined by two axes, the density of agricultural throughput per hour of labour and the density of agricultural throughput per hectare of land.

Considering the process of agricultural production, changes in AP_{ha} and AP_{hour} reflect changes in technical coefficients of technologies adopted at the farming system level. On the higher level, society as a whole, changes in AP_{DP} and AP_{hour} reflect alterations in the characteristics of the socioeconomic system to which the agricultural sector belongs. Being part of both levels, agriculture must operate in such a way that the combination of values of AP_{DP} and AP_{hour} is compatible with the constraints imposed by society, that is the combination of the values of AP_{DP} and AP_{SEP}. Clearly, farming systems of subsistence societies occupy tin area of the plane that is quite distant from that occupied by farming systems of industrialized countries. In general, farming systems tend to have an actual agricultural performance that is closely related to the set of socioeconomic constraints corresponding to the socioeconomic context in which they are operating. Describing changes in agricultural production on the plane formed by the agricultural throughput per hour of labour and the throuhput per hectare of land makes it possible to define trajectories of changes for both socioeconomic variables and technological coefficients in agriculture, which are expected to accompany trends in technological development. For instance, current trends in economic development suggest an economic growth of developing countries that is characterized by increases in both population size and per capita consumption. This will be reflected in a parallel increase in APDP and APSEP. Sooner or later, these changes in demographic and socioeconomic pressure (APDP and APSEP) will be reflected into altered technical coefficients of farming techniques. That is, technological development in agriculture in this situation tends to evolve toward higher values of AP_{DP} and AP_{hour}.

General trends in the evolution of food production techniques for different types of socioeconomic systems can be represented on the two-dimensional plane. Four main types of socioeconomic systems, having different combinations of demographic and socioeconomic pressure:

— *Socioeconomic systems with low demographic and low socioeconomic pressure.* This situation is characterized by an arable land per capita of more than 0.5 ha and a GNP per capita of less than US$10 000 per year.

This type of system would include several African countries, such as Burundi.

— Socioeconomic systems, such as the USA, Canada and Australia, with low demographic and high socioeconomic pressure, that is arable land per capita more than 0,5 ha, and a GNP of more than US$10000 per capita per year.

— Socioeconomic systems, such as PR China, and Egypt, with high demographic and low socioeconomic pressure, that is arable land per capita less than 0. 1 ha and GNP per capita less than US$1000 per year. Socioeconomic systems, such as the West European countries, with high demographic and high socioeconomic pressure, that is arable land per capita less than 0.2 ha, and GNP per capita more than US$10000 USD per year.

According to existing trends in population growth and economic development for these four different types of socioeconomic systems, we may expect the following movements in the plane:

— *Societies with low demographic and socioeconomic pressure.* The population is growing faster than the GNP per capita, which means that AP_{DP} will grow faster than APSEP. Hence, they will move toward a situation typical of PR China.

— *Societies with low demographic and high socioeconomic pressure.* Economic development is expected to be maintained and population growth will he relatively slow but steady. On the plane, this means a slow movement toward higher values of $AP_{DP.}$

— *Societies with high demographic and low socioeconomic pressure.* They tend to show economic growth and maintain if not expand their already huge population size, At national level, an increasing

GNP per capita will result in an accelerated absorption of the labour force currently engaged in agriculture by other sectors of the economy. This will inevitably require a major increase in agricultural labour productivity to maintain security. Hence, a movement toward the West European conditions of agricultural production is to be expected.

— *Societies with high demographic and socioeconomic pressure*. These societies have no alternative but to try to maintain a high material standard of living and keep population growth to a minimum. This means a more or less stable and high level of AP_{SEp} and a very slowly increasing value of AP_{DP}.

For these societies, trying to re duce the environmental impact of their food production becomes a major factor. Note that food import from the international market, a must for countries where $AP_{DP} > AP_{ha}$ and/or $AP_{SEP} > AP_{hour}$ is based on the existence of surpluses produced by countries where the relation between these parameters is inverse. At present, the USA, Canada, Australia and Argentina combined 80% of the net export of cereals on the world market.

However, at their present rate of population growth and because of an increasing concern for the environment this surplus may be seriously eroded ill the near future. For instance, the USA is expected to double its population in 60 years. Food import problems will be accelerated if a general concern for the environment will lead to a less intensive agriculture all over the world. At world level, demographic and socioeconomic pressure are certainly expected to increase. It is often overlooked that, at world level, there is no option to import food from elsewhere. When increases in demographic and socioeconomic pressure are not matched by an adequate increase in productivity of land and labour, food imports of the rich will be based on starvation of the poor.Worldwide, arable land per capita is already less than 0.27 ha per capita at present, and is expected to continue to shrink because of population growth.

In addition, arable land is being lost. For instance, during the past 40 years nearly one third of the world's cropland has been abandoned because of soil erosion and other types of degradation. Most of the added land that replaces this loss has come from marginal land made available by deforestation.

High productivity per hectare oil marginal lands requires large amounts of fossil energy-based inputs. This is one of the driving forces increasing the dependence of the world food supply on fossil energy. Many densely-populated countries are not able to meet their internal demand for food without heavy reliance oil fossil energy stocks. "Half of all peasants in Southern China are alive because of the urea cast ladled onto tiny fields—and very few of their children could be born and survive without spreading more of it in the years and decades ahead".

The increasing dependence of the food security of poor densely-populated countries on oil may become problematic in the near future. In the last few years. PR China and other Asian countries have been among the countries with the fastest growing economics in terms of GNP. At world level, fueling the economic development of densely-populated countries represents a major drain on natural resources, notably fresh water and fossil energy stocks. Not only do we risk that a dramatic surge in oil demand will augment oil prices, but we may soon face the bottom of the oil barrels.

The goal of achieving sustainability of food production in the future is closely related to the challenge of mediating between the socioeconomic constraints 'imposed' by society, that push for more concentrated and faster agricultural throughputs, and ecological constraints 'imposed' by natural processes that push in the opposite direction.

Natural processes provide the services on which all agricultural production relies, such as water and nutrient supply, soil fertility, pollination, stabilization of atmospheric composition, and regulation of biogeochemical cycles. Respecting ecological constraints means keeping the flows of biomass produced in the agroecosystem by human management within a range of values that is compatible with regulation mechanisms typical of natural ecosystems.

The mutual compatibility between socioeconomic and ecological constraints in agriculture is simply checked by comparing the minimum density of the throughput in the agroecosystem that would be needed to achieve, in an economically viable way, food security with the maximum densities that would still be compatible with the original ecosystem characteristics (keeping the environmental loading below critical thresholds).

A large discrepancy between the actual density of' agricultural throughput and the densities imposed by either socioeconomic or ecological constraints indicates unsustainability of agricultural production. Agriculture is ecologically unsustainable when the actual density of agricultural throughput exceeds the critical environmental loading, and socioeconomically unsustainable when the actual density of agricultural throughput is too low to provide food security and/or economic viability for farmers.

It is very possible that the actual agricultural throughput in many developing countries is both ecologically and socioeconomically unsustainable at the same time. The interface of agriculture with the larger socioeconomic system has been discussed in detail in the previous section. In this section, I explore the interface of agriculture with the larger ecological system.

Agriculture can be defined as a human activity that exploits natural processes and natural resources in order to obtain food and other products considered useful by society. The verb 'exploits' suggests that we deal with an alteration. Indeed, within a defined area humans alter the natural distribution of animal and plant populations in order to selectively increase the density of certain flows of biomass that they consider more useful for the socioeconomic system.

Adopting this definition, we implicitly assume that agriculture has to do with manipulation of biodiversity. Following the schema of ecosystem structure proposed by Odum, a natural ecosystem can be seen as a network of matter and energy flows in which nutrients are mainly recycled within the system and solar energy is used to sustain this matter cycling.

The amount of solar energy used for self-organization by the ecosystem is proportional to the speed of its matter cycles (in terrestrial ecosystems this also has to do with availability and circulation of water). Agriculture then involves a boosting of only those matter and energy flows in the network that humans consider beneficial, and an elimination or reduction of the flows that they consider detrimental to their purposes. In (Icing so, humans amplify some of the existing genetic information (that is contained in populations representing 'useful' biomass), and reduce or eliminate some other genetic information. In other words, changes in the hierarchical structure of matter and energy flows in the ecosystem must be coupled to changes in the hierarchical structure of genetic information clustered in species and distributed over taxa in the geographical space altered for agricultural purposes. The hierarchical structure of genetic information can be stabilized only by a quasi-steady-state in nutrient flows, which in turn depends on the ability to use available flows of free energy.

The resonance between coded information in DNA and the metabolism of living systems implies that the reverse must also he true. The particular hierarchical structure of matter and energy flows is determined by the rate of input/output of individual components of the network and their spatial and quantitative distribution; that is, on a large scale, it depends on the hierarchical organization of genetic information. Profound changes in the densities of harvested biomass can have dramatic consequences for the ecosystem structure. The elimination of ecosystem diversity can affects the stability of the biosphere when the scale of human alteration is increased to reach global relevance.

For example, in order to significantly increase yields of biomass (the amount taken away from the agroecosystem per unit of area and time), it is necessary to fight competing species (pests) and to bypass physiological rates of nutrient cycling in the ecosystem (fertilization) when these become limiting. This calls for selection of crop species and varieties that perform well tinder human-managed conditions.

The interaction of the agricultural sector with both ecosystem and society can be studied by following inputs and outputs of the four main steps of the food system cycle:

— Food production, where food is defined as energy and matter compatible with human metabolism.

— Making the food accessible to consumers, where accessible food is defined as meals ready to be consumed according to defined food consumption patterns.

— Consuming food and generating wastes, where wastes are defined as energy and matter no longer compatible with human needs.

— Recycling wastes to agricultural inputs, where agricultural inputs are defined as energy and matter compatible with the agricultural process.

All steps are interconnected, and hence the definition of a flow as an input, an available resource, an accessible resource, or waste, is arbitrary. First of all, the definition depends on the point of view from which the system is analysed. For example, the introduction of trees leads to increased evapotranspiration, which may be negative in terms of less accessible water in the soil, but positive in terms of more available water in the form of rain clouds. Second, the definition depends on the compatibility of the throughput density with the processes regulating the particular step in question.

For example, human excrements of a small Chinese village are a valuable input for agriculture, but the sewage of a big city is a major pollution problem. Whether or not the speed of a throughput at any particular step is compatible with the system as a whole depends on the internal organization of the system. Nobody expects science to make it possible to feed a person 30000 kcal of food per day, about 10 times the normal amount, while keeping that person alive for a long period of time. Why then do many believe it to be possible to increase the productivity of agroecosystems by several times without generating any, negative side-effects on agroecosystem health?

Indeed, the increasing intensity of agricultural throughputs over increasing areas of our planet should be considered a major threat to the conservation of biodiversity. The more the amount of agricultural biomass harvested from a defined population (considering a single species at the time) in an altered ecosystem differs in density from the biomass turnover of a similar species in naturally occurring ecosystems, the larger can be expected to be the environmental impact on the agroecosystem and the energetic cost of the agricultural production.

Indeed, in order to maintain in 'artificially' high density of energy and matter flow in the selected population of a certain species, humans must 'interfere' in the agroecosystem to amplify the genetic information of the selected populations and Suppress genetic information of competing populations. The first action must be backed up by a supply of the required inputs (external fertilizers and irrigation). The second requires active intervention in the life cycles of competing populations, such as slash and bum, tilling of soil, weeding, and pesticide applications.

Extremely high densities of agricultural throughput (per hectare and/or per hour) necessarily require production techniques that ignore the functional mechanism of natural ecosystems, that is the nutrient cycles powered by a linear flow of solar energy. Clearly, individual species, taken one at the time, cannot generate cycles in matter, since they perform only a defined 'job' (occupy a certain niche) in the ecosystem.

Technical progress sooner or later implies a switch from low-input agriculture based on nutrient cycling within the agroecosystem to high-input agriculture in which the throughput density of harvested biomass is directly controlled and maintained at elevated levels through reliance on external inputs that provide linear flows of both nutrients and energy. In low-input agriculture, where the harvested flow of biomass remains within the range of natural turnover values, the activity of production interferes only to a limited extent with the system of controls regulating matter and energy flows in the ecosystem. This form of agriculture requires that at least several distinct species are used in the process of production so as to maintain the internal cycling of natural inputs as a pillar of the agricultural production process.

For humans this implies a poor control over the flow of produced biomass in the agroecosystem, because of the low productivity per hectare when assessing the yield of a particular crop at the time and the variability in productivity in case of perturbations. Especially serious is the problem of facing dramatic increases of demographic pressure. This type of farming cannot work properly with levels of throughput per hectare too high.

On the positive side, direct and indirect biophysical costs of production are low. Few technical inputs are required per unit of output produced and, when population pressure is not too high, the environmental impact of this form of agricultural production is modest. The contrary is true for high-input agriculture. In this case, the harvested flow of biomass is well out of the range of values that is compatible with regulation processes typical of natural ecosystems (when considering yields of a single cultivated species at the time). In high-input agriculture, human management is based on an eradication of the natural structure of controls in the ecosystem.

In fact, where several tonnes of grain have to be produced per hectare and hundreds of kilograms of grain per hour of agricultural labour, natural rates of nutrient cycling and a natural structure of biological communities are unacceptable. In high-input agriculture, not even the, genetic material used for agricultural production is related to the original ecosystem in which production takes place.

In this way, humans keep the flow of produced biomass tightly under control and adjust yields to match increasing demographic and socioeconomic pressure (e.g. the green revolution). However, this control is paid for with a high energetic cost of food production, because humans must continuously provide artificial regulation in the agroecosystem in the form of inputs and defense against undesired, competitive species.

The environmental impact of high-input food production is necessarily large, and involves a dramatic reduction of biodiversity in the altered area, destruction of the entire set of mechanisms that regulated ecosystem functioning before alteration, and several negative side-effects such as on-site and off-site pesticide and fertilizer pollution, soil erosion, and salinization.

How do changes in technical coefficients of farming techniques, aimed at an increase of both land and labour productivity, affect biodiversity use in agriculture.' What is the nature of trade-off between biodiversity use M agriculture on the one hand and productivity of land and/or labour on the other hand? At the level of the farming system the parameters land productivity and labour productivity are interdependent. The existence of this link is illustrated by the possibility to express the parameter labour productivity as, the product of two terms, one of which is the productivity of the land itself:

$$\text{labour productivity (kg hr}^{-1}) = \text{land productivity (kg ha}^{-1}) \times \text{land managed per hour (ha hr}^{-1})$$

Hence labour productivity can be augmented by either

increasing the land area managed per hour of labour, increasing the land productivity or a combination of these two solutions. Usually, technical progress induces a combination of the two solutions which are mutually reinforcing.

The land managed per hour of labour can he augmented, while maintaining the same yield per hectare, by a substitution of machine power for human power. This solution pushes for monocultures, because a monocultures because a monoculture facilitates mechanization by, synchronizing and concentrating in time and space the requirement of inputs and the harvest of outputs. With this solution, labour productivity can be raised by using more land per farmer, while avoiding all increase in energy costs per hectare in the form of land-saving inputs such as fertilizers and irrigation. This solution is typical of farming systems in Canada and Australia, where the demographic pressure is low.

Since capital substitution for labour (human power) must have a positive economic return to pay back its own investment, economic feasibility is a crucial factor tor this solution. There are two obvious ways to increase farm revenues in this situation. First, by trying to get the most out of the local conditions by focusing on the crop(s) that show(s) the best comparative advantage in the area.

Second, by trying to get the most out of the capital investment by expanding the area of the farm, carefully concentrating cultivation in favourable spots and avoiding marginal lands. To facilitate mechanization, fanners generally tend to increase the dimension of their fields, by eliminating patchworks and edges, despite the fact that those landscape patterns and elements are known to be important in the maintenance of beneficial insect species for pest control and conservation of biodiversity in general.

Thus, the use of machine power to increase labour productivity will drive technological development in agriculture further toward the exploitation of comparative advantages and economies of scale (e.g. enlargement of farms). The building of' specialized technical infrastructures in the area will reinforce the trend of product specialization and generate a mode-locking effect.

While increasing yields per hectare without enlarging the land area managed per unit of labour can be a (temporary) solution to match an increasing demographic pressure, it has limited power to meet the demand for a higher labour productivity. The problems with this solution are evident in PR China, where massive fertilization and irrigation are practiced to augment yields coupled to labour-intensive farming techniques. Although in this way agriculture will feed a densely-populated country, it will provide little support for its socioeconomic development.

The equation of labour productivity indicates that both an increase in land productivity and an increase in land managed per hour of labour can have a positive marginal return in terms of an increased labour productivity and economic return. Therefore, the particular choice of technical inputs in agricultural production—machinery to boost the land managed per unit of labour or fertilizers and irrigation to boost land productivity—will depend on the expected marginal return of capital investments in these two factors. Sparsely-populated developed countries, such as Canada, the USA and Australia, tend to invest more in machinery, even though a certain investment in land-saving inputs is made when the marginal return is high.

Densely-populated developing countries, such as PR China, Egypt and Bangladesh, tend to predominantly invest in fertilizers and irrigation, and invest only little in machinery. Densely-populated wealthy countries, such as Japan, Italy, and the Netherlands, are in a peculiar situation. In these countries, farming can become a 'mission impossible', because of the simultaneous high demographic (arable land per farmer cannot be significantly expanded) and socioeconomic pressure (the need for high labour productivity and economic return in the agricultural sector).

Despite heavy investments in both machinery and land-saving inputs, farmers in these countries still face the unpleasant reality of a land and labour productivity that do not match the minimum outputs imposed by their demographic and socioeconomic pressure.

For instance, it is impossible for a Japanese farmer to produce hundreds of kilograms of grain per hour of labour by working 1800 hours per year on just one hectare of arable land. When biophysical constraints of this type exist there are no other Ways out than either resorting to a system of government subsidies (transferring money from other economic sectors to agriculture) to guarantee acceptable farmers' income (comparable to that obtained in other economic sectors) and/or importing food to guarantee food security to the country.

The need for high labour productivity (socio-economic pressure) and eventually land productivity (demographic pressure) at the level of society, reflected into changes in technical coefficients at the farm level, is intricately linked to changes occurring in the energy budget of socioeconomic systems to which the agricultural system belongs. Economic development, assessed at the socioeconomic level, inevitably results in a continuously increasing opportunity cost of human labour and capital at the farm level, and pushes technical development of the farming system towards mechanization, specialization, economies of scale, linearization of nutrient inputs in the agroecosystem, and dependence on price support and other subsidies.

This chain of changes, well-known in the field of agricultural economics though generally described from a different perspective, is the main cause of the abandonment of traditional farming strategies, and the consequent loss of biodiversity in agriculture. Most notably, two important trends follow from this process of technological change: adoption of monocultures, and reduced use of by-products, such as crop residues and manure, in the agricultural process.

Natural ecosystems (e.g. mature natural forest) are able to sustain a large amount of biomass, but with a net primary productivity that is relatively low compared to agroecosystems. Traditional farming systems based on gardening in natural forests can produce large quantities of biomass per hectare on the cleared areas, but only with a long cycle of rotation. When the entire area involved iii the rotation is considered, overall crop productivity per hectare is quite low.

Other systems of traditional farming based on inter-cropping, multiple cropping and animal husbandry have proved to be highly productive in terms of food supply per hectare whenever ample fresh water is available. How ever, the combined exploitation of diverse natural resources generally requires labour-intensive manage merit. Rural communities in China and elsewhere in East Asia, masters in the integrated use of natural resources, have been able to feed themselves and pay taxes in form of part of the harvest in spite of a limited endowment of land, for centuries.

However, the particular techniques of multiple cropping, animal husbandry and integrated use of natural re sources implied that more than 80% of the population was engaged in agriculture. Moreover, note that the population of China did not surpass 600 million until synthetic fertilizers were made available to China m the late 1970s. Whenever socioeconomic development calls for a reduction of the labour force in agriculture, the society must confront the need for an increase in farm labour productivity.

For example, the yearly amount of food consumed by a US citizen is produced with less than 20 h of farm labour. Up to the present, this level of agricultural labour productivity has only been achieved by making use of monoculture. Mounting demographic and socioeconomic pressure can make traditional techniques, such as combined production of cereals and legumes, and rotation with green manure, no longer feasible unless *ad hoc* policies are established to provide government support

Recycling of natural inputs within the farming system is abandoned as soon as this becomes too costly in terms of land and/or human labour requirement, provided of course that alternative technical inputs for use in agriculture are available on the market For example, the trend of abandoning cultivation of green manure in favour of food grain when artificial fertilizer is accessible is discussed in detail by Smil. In general, the decision whether or not to abandon the recycling of natural inputs will depend on the amount of energy or nutrients gained by recycling, divided by the extra hours of labour required, the opportunity-cost of labour, and the price of technical inputs used as alternatives.

When the inputs made available per hour of labour by recycling are significantly less than the inputs that could be bought on the market with the economic return of one hour of labour allocated to off-farm activities, then recycling will generally be abandoned. In this case, linearization of nutrient inputs is more convenient in terms of economic cost/benefit. Within a market economy, the choice to specialize and rely on monocultures often leaves no alternative but to also switch to linearization of nutrient inputs in the form of fertilization.

The continuous and fast changes in the socioeconomic context of farming are driving technological progress in agriculture toward solutions (densities of agricultural throughputs) that are increasingly incompatible with ecological processes. The need to keep pace with mounting demographic and socioeconomic pressure is affecting agricultural development all over the world, and threatens biodiversity in both wild ecosystems (their area is shrinking) and managed ecosystems (the intensity of alteration is increasing).

Criteria adopted for choosing and developing farming techniques have generally been based on socioeconomic constraints only (costs/benefits referring to economic growth) and therefore have been biased in favour of continuously increasing densities of harvested biomass.

Ecological constraints imposed by the characteristics of natural ecosystems, that would put an upper limit to the densities of harvested biomass, tend to be ignored. This is for the simple reason that socioeconomic constraints to agriculture leading to technological 'progress' have universal application (technical coefficients refer to unit operations in artificial environments and can easily he generalized in terms of monetary costs/ benefits), whereas ecological constraints are location-specific and extremely difficult to define and quantify. The consequent imbalance between the short-term and long-term (ecological) perspective when deciding how to manage agricultural production may have disastrous consequences given the immense scale of agricultural activities worldwide approximately 50% of the terrestrial area is devoted to agriculture, while in temperate ecosystems agriculture occupies 70% of the land.

As Hollings has pointed out, in biological systems a general homogenization of activities over a large scale is a disaster waiting to happen: it dramatically increases the risk that a perturbation damaging to that pattern will occur and be amplified over the whole system.

In conclusion, strict respect the pattern of natural process in the agroecosystem to preserve biodiversity would slow down the throughput of food production per hectare of land and per hour of labour. When these throughputs become so low so as to become incompatible with socioeconomic characteristics this solution becomes economically non-viable.

On the other hand, matching demographic and socioeconomic pressure above certain ecological thresholds leads to ecological unsustainability. For example, the production of tonnes of grain per hectare per year and hundreds of kilograms of grain per hour of labour with less than 0.2 hectare of arable land per capita can only be obtained in an ecologically unsustainable way through linearization of nutrient flows in the agroecosystem, that is continuous harvests (no rotation) and heavily reliance on technical inputs such as machinery, fertilizer, pesticides, and irrigation.

Discussions of sustainability of world agricultural production should carefully consider both ecological constraints such as the possible limiting role of land, water, and other natural resources, and socioeconomic constraints to food production.

Current strategies of technological development in agriculture not only appear unable to solve problems of food security in the long term, but also have been increasingly malfunctioning in the last decade, both in developed and in many developing countries.

In developed countries this is because excessive socioeconomic pressure has caused farms and fanners to disappear, while remaining farming systems employ production techniques that are detrimental to ecological equilibria, and in developing countries it is because excessive demographic pressure has caused poverty in rural areas to increase and food production per capita to decrease.

Paradoxically, this evident malfunctioning is perhaps also a reason for optimism. After all, when dealing with human systems, cultural mode-locking is the most persistent and important cause of lack of adaptability in response to changing boundary conditions.

As concerns agriculture, however, there is an increasing, though not yet general, consensus among farmers, scientists, politicians and the general public that we somehow have to change the way we look at and practice agriculture. In terms of opportunities for future changes in farming strategies, the situation is dramatically different in developed and developing countries.

In developed countries, where high-input agriculture is the rule rather than the exception, there is some room for maneuvering to get out of the current path of development. Intensive agriculture has been driven by a strong and fast socioeconomic development and based on the creation of a 'risk-free environment' for farmers and investors in the agro-industry.

This risk-free environment, that was and still is financially supported by the rest of society, has been realized through guaranteed prices, input subsidies, disaster, relief, etc. Indeed, modem high-input agriculture depends economically on the rest of society, that mobilizes economic resources to maintain its agricultural sector. We may thus conclude that in developed countries the economic resources are available to support the agricultural sector, but that they should be spent in a different way.

Developing countries have never disposed over sufficient economic resources to generate a risk-free environment for farmers and investors in high-input agriculture. On the contrary, the agricultural sector is usually taxed to support other sectors of the economy.

In the absence of government support, risks implied by excessive simplification of ecosystems typical of high-input agriculture are transferred to individual farmer families. This makes such a solution less attractive, especially for poor farmers operating in marginal areas. Nevertheless, the continuous increase in demographic and/or socioeconomic pressure as well as access to Western technologies, has pushed for the adoption of high-input techniques in local attempts to augment agricultural throughputs. Given the rapidly growing population in developing countries (actual increases in demographic pressure) and the gap still to fill before reaching desired income levels (potential socioeconomic pressure), we can expect the density of agricultural throughputs to increase in most, if not all, developing countries, even though this could be destructive for the environment in many cases.

In spite of this wide difference, there is a striking similarity between high-input and low-input agriculture, In both cases, food security totally depends on the availability of fresh water, fertile soil, pollination, natural recycling of nutrients and pollutants, biological pest control, and nitrogen fixation. These are services that cannot be substituted by injection of technical capital, and that are heavily dependent on biodiversity.

The neoclassic economic paradigm endorses the idea that unlimited trade is a useful tool to enhance world economic growth. Adopting this tool, individual countries can overcome biophysical constraints in the form of local shortages of natural resources that limit the expansion of their economics. According to this view, trade enables a better use of natural resources through complementation of economies (comparative advantages).

However, this view implicitly implies the adoption of the entire world as the hierarchical level at which the economic process is assessed and optimized. This level is not necessarily the best possible choice, especially when dealing with agricultural production. In fact, the use of a smaller space-time scale for defining strategies of technological development also has positive effects. First, a smaller scale makes it easier to involve local people in the decision-making process and increase the responsibility of local communities for resource management.

Thus, decisions related to sustainability should be made as close as possible to local people, and with the participation of all major stake-holders. Apart from the ethical aspect, this would make it possible to benefit from indigenous knowledge of the nature of the problem of sustainability dealt with. Second, internalizing most of the external services on which economic systems rely (by reducing trade) makes it easier for the system to respond quickly to changes and signals coming from surrounding ecosystems, in spite of their intricate and varying nature. Societies heavily relying on trade often cannot see the importance of the ecological services on which they directly and indirectly rely. Therefore, energy expenditure in transportation from distant ecosystems should be justified only in cases of substantial comparative advantages. Third, the goal of harmonizing energy and material circulation with the characteristics of local ecosystems (decreasing the environmental impact of' economic activities) can he more easily achieved it' thc spacc-timc closure of the activities of production and consumption in the society is similar to the space-time closure of the matter cycles occurring in local ecosystems. These points suggest that the enlargement of scale of economic systems obtained by heavy trading can also induce negative effects, both on the stability of the biosphere and on the integrity of local socioeconomic systems.There is therefore also a rationale for policies aiming at increasing the level of self-sufficiency and addressing the issue of social equity at the level of national and regional economic systems.

When defining the terms of reference for sustainable agriculturc, biodiversity should be considered as an output rather than as an input in agriculture. Agriculture should not only produce food, but also preserve a healthy environment. The latter task, grossly neglected, involves guaranteeing a higher level of cycling of nutrients, better control of pollution of fresh water supplies, and a sound management of the landscape in general. For all these tasks preservation of biodiversity is csscntial.

Conservation of biodiversity, however, only makes sense within a large-scale planning, both in terms of time and space. For instance, if socioeconomic pressure imposes high-input agriculture, then we should exploit only those terrestrial ecosystems that are better adapted to being altered, including rotation and fallow.

Improving long-term sustainability of agricultural production implies more emphasis on ecological constraints, even if this implies reducing the short-term efficiency of agricultural production. Indeed, if the 510 year average discount period typically adopted by economic analyses is not abandoned, then there is little chance that assessments of costs and benefits referring to biodiversity will be seriously considered.

Moreover, the perception of the trade-off between long-term ecosystem health and short-term maximization of production is affected by the level of development. In countries with a high material standard of living, citizens often consider it acceptable to reduce economic returns for the benefit of the environment, while developing countries do not have room to reduce economic returns. They have no other option but to boost short-term returns at the cost of long-term sustainability.

At the moment of discussing agricultural policies in view of sustainability, the small space-time scale typical of economic analyses should always be balanced by a long-term and worldwide view. But is this a feasible idea? Developed countries might be prepared to sacrifice part of their own economic return for the benefit of their immediate surroundings, but doing this for the benefit of other, distant societies and ecosystems is something quite different. The fact that all socioeconomic systems, ecosystems and agroecosystems form part of a larger whole, the functioning of which is affected by the health of all its individual components, often appears to be beyond our capacity and willingness of understanding. The 'biodiversity revolution in agriculture' should result in a definition of a set of new optimizing parameters for agricultural techniques.

9 Threat of Biodiversity Loss

Although tropical forests indeed have the highest biodiversity in terms of total number of taxa, and number of taxa per unit of area, their cover worldwide is giving way to timber industries, mining, advancing agricultural frontiers, and other causes, at a rate of about 1% per year Active conservation of biodiversity is still largely confined to protected areas, but those cover only a little more than 3% of the world's land area.

Realizing these trends, many biological scientists have expressed the need for conservation of biodiversity in land areas other than natural tropical forest. Data from the last three decades demonstrate that on a worldwide scale the total area under forest and woodland has remained fairly stable.

Considering the conversion of natural forests in many parts of the world, this means that in general nature natural forests are being replaced by woodlands which are the result of human activity, be it planting of trees or the development of secondary forests on land which has been cultivated in the past. With the increasing demand for land, and for wood worldwide, this trend is likely to continue for decades to come. Consequently, it can be expected that forests which bear the marks of human influence will play an increasingly important role in species conservation, particularly in tropical countries where the largest biodiversity must be preserved, but also where economic needs put considerable pressure on natural resources. This is being increasingly recognized by forestry officials and forest managers worldwide.

Many countries in the temperate zones which no longer have any significant amount of mature natural forest manage biodiversity in anthropogenic forests. In many European countries there is only little need to generate economic returns from those forests, while in the USA the National Forest Service recognizes a serious need to balance economic demands with biodiversity.

In many tropical countries where forests are managed for economic interests, maintaining a diverse forest with a large number of species is often considered contradictory to the principal goal of increasing returns. This, for instance, remains the case with most of the plantation and land development programs that the Indonesian Ministry of Forestry has designed for areas of forest land which nave been designated Conversion Forest. Indonesia has a total land area of 199 X 10^6 ha, including about 109 X 106 ha of forest and woodland of which 86.4 X 10^6 ha are forest cover. This is about 6% of the world's total remaining tropical forest, and the third largest area after. Brazil and Zaire. Indonesia's forests contain about 10% of the world's plant species, 12% of the world's mammals, 16% of the world's reptiles, and 17% of the world's bird species. Because of its unique geographic location on both sides of the Wallace Line, between the continents of Asia and Australia, a high rate of species endemism occurs.

On the other hand, Indonesia's forests constitute an important economic resource for the country. Until recently, timber was the second largest generator of foreign exchange after oil, and remains an important source of development capital Not with standing this economic importance, Indonesia has repeatedly expressed an awareness of it's responsibility to the world community to preserve it's forests and the biodiversity they contain. In order to safeguard both the important economic role of the timber industry and the country's biodiversity, Indonesia has functionally divided its large forest. This division was laid down, until 1996, in the Forest Land Use by Consensus which is now being replaced by the Provincial Land Use Ordination Plan. Roughly one-third of the country's forest lands are set aside for conservation or protection purposes.

The larger part is designated as Production Forest, while the remaining area is designated as Conversion Forest or Nonclassified Forests. Conversion and Non-classified Forests may be used for agricultural or other purposes. The total of the previous forest land categories is 144 X 106 ha, a result of the fact that not all forest land actually has forest.

The Government of Indonesia considers one of the major threats to the timber-rich forests of Indonesia to be the many swidden agriculture farmers in the outer islands where the larger areas of Indonesia's forests remain. Hence, the GOI has made it a goal to eradicate all forms of swidden agriculture in the whole country. It has designed several programs to reach this goal. These programs vary according to the category of forest in which they are being conducted.

So called 'social forestry' programs have been designed to stop forest encroachment by small farmers in Conservation, Protection, or Production Forest areas. In forest land that is designated as Conversion Forest, swidden agriculturists are encouraged to participate in estate crop development programs, especially rubber and oil palm, or to become wage laborers in commercial timber estates. In this chapter we will discuss the programs that the GOI has been developing for areas of Conversion Forest, and juxtapose them to a case of traditional forest management by swidden farmers in West Kalimantan, one of the four Indonesian provinces in Borneo.

The four provinces of Kalimantan hold 44 X 10^6 ha of forest land, or 31% of the total forest area of Indonesia. In 1990, 34 X 10^6 ha still had mature standing forest. 11 X 10^6 ha are permanent Conservation and Protection Forest, 26 X 106 ha Production Forest, and 8 X 10^6 ha Conversion Forest. A total of 4.4 X 10^6 ha are estimated to be under swidden agriculture, which sustain 228 000 households, or 17.2% of the total population. The forest cover in Kalimantan is estimated to disappear at a rate of 0.49% $year^{-1}$.With the present discussion, we hope to demonstrate that the programs designed to reduce the threat that swidden agriculture supposedly poses for In donesia's forests will necessarily lead to a significant reduction of the high biodiversity typical of many areas which are, predominantly, under swidden agriculture.

Furthermore, we argue that alternative choices for economic development of swidden agriculturists can be designed which can preserve, or even enhance the biodiversity in these same areas. In particular we will try to demonstrate: (1) that conservation of biodiversity is an integral part of resource management among the very same swidden agriculturists as an important subsistence strategy; (2) that several of the programs which are designed to eradicate swidden agriculture in Indonesia ironically will lead necessarily to a loss of this biodiversity that is maintained by swidden agriculturists; and (3) that instead of promoting large-scale timber plantation or estate plantations, developing traditional forest management is both an attractive option for regional economic development as well as a way of preserving the biodiversity that is an integral part of many swidden agricultural landscapes.

It is estimated that swidden agriculturists are responsible for about half of the estimated 1 X 106 ha land deforested in Indonesia yearly, although little accurate data exist to sustain such an estimate. Of the total area under swidden agriculture, about 25% is in Production Forests, 15% in Protection Forests, 29% in Conversion Forests, and 31% in Unclassified Forest. Even though 60% of swidden agriculture is located in forest land which is planned to be used for non-forest purposes, swidden agriculture is considered to have serious implications for the future of forestry in Indonesia. Since the early 1970s, the GOI has designed programs to deal with swidden agriculturists. In their earlier phase these programs attempted to relocate swidden agriculturists to out of forest areas.

Especially since the mid-1980s. when the issue of' sustainable forest utilization has become widely discussed in Indonesia, many directives from the Ministries of Forestry, Agriculture, Transmigration, Social Affairs and Home Affairs have appeared which are meant to 'stabilize' swidden agriculture. The social forestry, or village development programs, which are applied in Conservation, Protection, or Production Forest areas, aim to give people responsibility for managing the forests in their areas. In fact social forestry programs are still ill-defined, and subject to much discussion. Experiences with social forestry in Java, where local control over forests was replaced by government control some time ago, give little hope for similar programs for the outer islands. Social forestry is not attempted in areas which have been designated as Conversion Forest. In this category of state forest land three different programs have been defined: (1) the establishment of industrial timber estates ; (2) the nuclear estate program and (3) the Rehabilitation and Expansion of Export Crops program.

The timber estates development program began in 1984. This program has a high, priority with the GOI, since in the near future, it is supposed to substitute for the expected decline in supply of timber as a result of exhaustion of the natural forests. Until 1988, some 1.4 X 106 ha of timber production plantation existed. The largest parts of those timber estates are still found in Java, but the second largest parts are located in the four provinces of Kalimantan with most in West Kalimantan Indonesia's goal is to reach 6 X 10^6 ha of industrial timber plantation, by the year 2000. These plantations are to be established in Conversion Forest or Production Forest areas if they do not fall under a forest concession and do not contain sufficient commercial natural forest. In practice many state HTI projects have been established in secondary forest, scrub land or grass land. The location of HTI estates in secondary forest areas implicitly means that it is located in land that is occupied by swidden agriculturists.

One of the objectives of HTI estates is also to rehabilitate degraded lands. However, often any land that is under fallow is qualified as degraded by forestry officials, although from the swidden agriculturists point of view this land still may be productive rice land. HTI planners give little consideration to the actual importance of this rice land for swidden agriculturists. The HTI estates are largely to be planted with a few fast growing species. For instance, INHUTANI III the state company in charge of establishing the industrial forest plantations in West Kalimantan, has been planting only *Pinus merkusii* in the HTI estate in the district of Sintang, in which eventually 100 000 ha of land are to be planted. In the neighboring district of Sanggau about three times !bat area are to be planted to *Acacia mangium.* The participation of swidden agriculturists in industrial timber estates is mainly as providers of labor and land. Land needed for the estates is obtained from local farmers who are given some compensation, but this is not higher than US$15.00 ha1.

In the original wans the timber estates were to provide work to the local population, hence generate income, and thus reduce the need for swidden farmers to grow their own food, and slash more forests. The wages paid are rupiahs 3750 day 1, or US$1.70. Only when the implementation of industrial timber estates is combined with the Transmigration Program are housing infrastructure, schools, medical posts, and the like, provided to the transmigrants coming from Java, Madura, Bali, and Lombok, as well as local farmers who agree to participate. HTI/TRANS projects aim to include about 50% of 'local' transmigrants in their schemes.

The NES/PIR or Nucleus Agricultural Estate Program, which began in 1979, and which is the responsibility of the Ministry of Agriculture, aims at establishing estate crops. In this program government or private companies establish an estate, which may be as large as 50 000ha. About 60% to 80% of the land is dedicated to smallholder production of the estate crop. Participant farmers are given 2 ha of land to establish the estate crop, and 0.5 1 ha for a house lot and a home garden. The estate company provides facilities to the participating farmers to establish their 2 ha of estate crop, and it guarantees to buy the yield, as well as to provide technical assistance and credits for inputs.. Farmers accept these facilities as a loan which they have to pay back. In West Kalimantan farmers must repay US$4900 plus an interest of 6%. Farmers who become estate members commit themselves to work as laborers on the company-owned part of the estate, for which they receive daily wages. Incomes that participating farmers receive from the sale of oil palm fruits are supposed to climb from RP37 500 month' after 3 years to RP 220 000month1 after 12 years. These predictions, however, do not agree with recent actual earnings after 6 years of only RP88 000month1 that farmers in some Nuclear Estate Programs obtained.

A new version of the same program is the **PIR/** TRANS program. This also establishes nuclear estate crops, but largely with the participation of transmigrants coming mainly from Java. Bali and Lombok. In many PIR/TRANS projects local swidden agriculturists are incorporated also, but since the estates rely on transmigrants for their labor, fewer local farmers can be involved which means that relatively more farmers will be resettled. Farmers usually do receive compensation for the land they give up.

Reports, however, document the coercive, methods used by the state companies to obtain traditionally held land. NES/PIR or PIR/TRANS estates are entirely converted to mono-crop plantations, except for a smaller part which is converted to food crop area. The plantations used to be slash-weeded frequently, but increasingly chemical defoliants such as Monsanto, Roundup and Kleenup are used, and under the mono-crop only a grassy weed vegetation remains. Hence, the claim that NES/PIR or PIR/TRANS projects aim to conserve the environment contrasts sharply with their actual outcomes.

In West Kalimantan, the larger part of NES/PIR or PIR/TRANS plantations are under African oil Palm and smaller areas under rubber plantations. In 1993 the total area under oil palm plantation in West kalimantan was 142 403 ha The total area planned to be converted into nuclear estate crop is 1.2 X 106 ha, or 8% of the total surface of West Kalimantan.

The third program that affects swidden agriculturists in West Kalimantan is the Rehabilitation and Expansion of Export Crops program. This program provides credit to farmers to improve the quality of their smallholder plantations. In the outer islands, support is given for 2 ha of export crop land. In West Kalimantan this program only focuses on improving the quality of rubber production.

Very soon after, the first *Hevea* rubber appeared in Sarawak early this century, it also was brought to West Kalimantan by the Dutch. Most rubber, however, has been grown as 'forest rubber', planted on swidden fields, and left to grow, together with the spontaneously, occurring secondary vegetation. As a result, most rubber gardens are diverse, species rich, anthropogenic forests

Farmers who receive credit and technical assistance under the Expansion of Export Crops Program commit themselves to plant only improved rubber clones. These are high yielding rubber clones, which are much less resistant to pests than the locally used varieties. In many cases, how ever, farmers have actually been sold the locally existing rubber as improved high yielding rubber. Income levels again are much lower than predicted when farmers where invited to participate in the project.

The numbers of trees on single improved rubber holdings varies between 160 and 800 trees with the highest number of 800 trees, after 9 years an income of Rp 120000 month $^{-1}$ can be achieved. This is significantly above the average rural cash income in the region, but must cover purchase of most household provisions.

Farmers who plant improved rubber clones are supposed to conduct a thorough slash-weeding from the onset of the plantation until the trees are large enough to be tapped. As a consequence, improved rubber gardens have a characteristic mono-crop species composition which is similar to that of Pinus *merkusii*, *Acacia mangium*, or oil palm plantations and not like the diverse forest rubber gardens found in Kalimantan and Sumatra. A dominant, mono-culture plantation has a grassy weed vegetation covering the forest floor. Since currently more than 400000ha of land are under traditional species-rich smallholder rubber gardens in West Kalimantan, the extension of loss of biodiversity can expected to be considerable if these areas are to be converted to monoculture improved rubber gardens.

The island of Borneo, with an area of 642 000 km2 consists of the two East Malaysian states Sarawak and Sabah, and the four Indonesian provinces of East, South, West, and Central Kalimantan. In 1990 the whole island had 12.7 X 106 inhabitants, 9.1 X 106 of which lived in the Kalimantan provinces. About one-third of the people who live on the island are broadly classified as indigenous Dayak.

Throughout the history of Borneo the island has been of interest for its natural resources. In pre-Dutch colonial periods for various forest resources and minerals, and for the last several decades for its oil and timber. On the other hand, agricultural products such as pepper and rubber have been important export crops since about the turn of the century. Production of oil palm also has been increased drastically in the last decades.

The changing economic importance of the island also has resulted in considerable social changes. In the Indonesian part, until two centuries ago, it was mainly Malay who inhabited the coasts and river banks while indigenous inhabitants lived in the interiors. Gradually various other groups came to claim their niche: the

Chinese, the Bugies from Sulawesi, and later people from other islands from the archipelago, as transmigrants, or as spontaneous migrants, continuously pushing back the native Dayak to more remote places. Together with the booming of timber and oil exploitation came the large scale opening up of large parts of West, South and, to a lesser extent East Kalimantan. This opening up was followed by large scale agricultural development projects. The native Dayak population mostly living in the remoter interiors relies for its daily staple on upland rice cultivation, produced in yearly cleared swiddens. Much controversy and confusion still exists about swidden agriculture. The term covers agricultural practices which begin with the slashing and burning of vegetation of an area of land, using it for intensive cultivation for one or several years, followed by fallowing the same plot for a longer period.

Among most swidden agriculturists, fallow fields continue to have an economic importance since the secondary vegetation contains spontaneously occurring species, or perennial species which, were planted at an earlier stage and which continue to yield useful products. Some of these species occur mainly in fallow vegetation, and are only exceptionally found in primary forest Jessup claims that some Kenyah farmers in East Kalimantan purposely maintain a sufficient area of fallow to assure the supply of such species.

Hence, as a rule, among swidden agriculturists some kind of management of the fallow vegetation occurs. Among many swidden agriculturists the dominant economic imperative is to optimize returns of labor, rather than returns of land. In order to achieve optimal returns per laborer, swidden agriculturists prefer to fallow their land, since this will avoid low yields front plots, or excessive need for labor for weeding.

Therefore, although fallowing of agricultural land is not limited to the tropical regions, swidden agriculture is an adequate adaptation, especially to these hemispheres with naturally low soil fertility, and high rates of weed invasion due to a hot and moist climate, as long as land pressure remains low. Swidden agriculture is still practised by many farmers in the world, and subsequently appears in,many different forms. Even on the island of Borneo, significant variations of swidden agricultural patterns can be observed. For instance, lban farmers historically had the reputation of preferring land that was newly opened in mature natural forests, and shied away from making swiddens in secondary forest, although Jessup and Sellato question the validity of such generalizations based on ethnic affiliation.

Other groups, like the *Kenyah and Kenyah,* cultivated the area around a new settlement for several decades, and then moved the whole village to a new site. These farmers cultivated land only a few times, as long as they stayed in one place, leaving it to recover for long periods after they moved.

Often new settlements were constructed within less than one day's walking distance from the previous sites such that old fallows or forest gardens could easily be harvested when fruiting occurred. A reason to move the village could be the exhaustion of agricultural land, or because diseases occurred at increasing frequencies, or in some cases the location had become unsafe as a result of invasions by other groups. Recent migrations have occurred because people wanted to gain easier access to markets and government services.

The *Bidayuh,* a name originally used to distinguish the non-Malay, non-Iban inhabitants of Sarawak's first division, constitute several different linguistic subgroups, who originated from the Indonesian part of West

Kalimantan, and Which are found as far South as the Ketapang basin. Bidayuh farmers practice a type of swidden agriculture which is more sedentary than that of several other groups in Borneo who migrate more often. Rather than moving entire villages, Bidayuh villages which become too large, are split and only portions moved to a new area of forest to start their own settlement. The stable Bidayuh villages, however, frequently move the location of their settlement within the village territory, over distances that do not exceed about 1 km. The more permanent character of Bidayuh settlements has certain consequences for the way swidden agriculture is practised. In order to be able to reside over the long term in a single region, a farmer must have a sufficiently large holding of land under various stages of fallow. This holding is built up gradually from when a couple become an independent farming household, until the area is large enough such that fields can be used and fallowed at sufficiently great intervals.

Since *Bidavuh* farmers have more stable holdings, the area to which they hold tenure claims are smaller than, for instance, the ever-moving *Than.* On the other hand, land tenure claims are strong, and are held by descendants for many generations, even after a settlements splits and one part of the community moves somewhere else. Another important feature which is typical of *Bidayuh* sedentary swidden agriculture is the emphasis on tree planting and conservation of forest resources.

Although *Bidayuh* swidden agriculturists engage in frequent conversion of species-rich mature natural forest into species-poor agricultural land in terms of higher plants and larger animals, their resource management, in general, is very conservation minded. Honorable as this may sound, this is mostly a result of economic interest. The landscape of a typical *Bidavuh* village is usually characterized not only by land under swidden and fallow, but also by many patches of forests of various sizes. For instance, among two different *Bidavuh* groups in the North of West Kalimantan, *Mate* Dayak and *Jangkang* Dayak, six types of managed forests have been described. These include communally held forest, or *hutan tutupan,* which are usually located close to villages and may be as large as 100ha.

Hutan tutupan mostly function as a reserve for timber for house construction, rattans, medicinal plants, fishing and hunting sites, and to protect water sources. A smaller version of the same type of forest, of between 1 and 3 ha, is forest islands *or pulau rimba.* Such patches of forest become surrounded by one household's agricultural fields or fallows and according to local customary law, became private property of this household.

These forests are often kept as private forest reserves, and yield the same products as *hutan tutupan.* In the same category of protected areas of mature natural forest fall the *sompuat,* which are patches of forest which surround honey trees, mostly *Koompassia excelsa* or *Koompassia malaccensis.* The *sompuat* forests are kept intact in order to provide an appropriate habitat for the *Apis dorsata* honey bees that built their nests in the central tree. Obviously, preserving areas of mature natural forests also implies preserving the diversity of species that are present in them.

However, that is not where *Bidayuh* biodiversity forest conservation stops. Two types of managed forests are important, riot only economically, but also for the diversity of plant species they contain. These are *tembawang* or fruit forest gardens, and rubber gardens created by individual farmers planting trees on sites previously used for agriculture or around the settlement area, or sometimes directly in mature forest. Planting may continue for many years until eventually most of the area designated for tree planting is covered. Such *tembawang* are privately owned and inherited by descendants of the people who established the forest. They are usually about 1-2ha in size.

Since many farmers in a single village create individual *tembawang* on adjacent fields, over time these merge into a single stretch of contiguous forest. Because of the short distances over which the village is frequently moved, the *tembawang* area often trails the advancing settlement as a long band of high closed forest. Rubber gardens are usually located adjacent to *tembawang.*

Once the rubber seedlings are planted, no weeding occurs until the rubber is fully grown; about 10 years. Only then are trails established to reach trees which have to be tapped. Very often additional fruit trees are planted in rubber gardens, as a result of which they tend to become an extension of the *tembawang* area, rather than a totally different type of managed forest.

The biodiversity conservation role of managed natural forests like *hutan tutupan, pulau rimba,* and sompuat are obvious. However, in this respect tembawang and rubber gardens are no less important. Although the principal products planted in a tembawang are fruits, bamboo, rattans, ironwood illipe nut or other dipterocarp species, many other naturally occurring species appear spontaneously in tembawang forest.

Often, if they are of economic interest, these species are protected or tended, while other species are tolerated and harvested as needed. Mature tembawang are full grown forests, which have a structure that is similar to that of mature natural forest and which hold a considerable biodiversity. In one of the villages researched, for instance, 400 plant species were recorded in five *tembawang*. These five *tembawang* had an average of 170 plant species per hectare, while one specific *tembawang* had a total of 231 plant species per hectare.

In two plots of mature natural forests in the research area 256 and 258 plant species per hectare were recorded. High species diversity can also be found in rubber gardens. The proportion of such managed forests in the landscape further stresses their importance for the conservation of biodiversity. In one village we found that of the approximately 1800ha under cultivation 600ha are tinder species-rich tembawang or rubber gardens.

It should be evident that any of the three programs that the GOI wants to implement in Conversion Forest areas will necessarily result in a considerable reduction of diversity in the landscape and the biodiversity it contains. From the biological conservation standpoint therefore, it would be better to leave swidden agricultural land the way it is, rather than try to implement large-scale plantations of mostly single species.

However, the objectives of HTI, NES/PIR and NES/TRANS and export crop development are not primarily to conserve biodiversity, but rather to increase or sustain the state's income from timber production, or estate crops. These incomes are indeed intended to increase living standards of local fanners, but also enhance the country's timber, oil palm and rubber industry, in order to provide income for ·the state and to generate foreign exchange.

The programs that deal with swidden agriculturists in Production, Protection, or Conservation forests, focus largely on developing alternatives which reduce the reliance of farmers on forests as a resource to create agricultural land. Some of the strategies are to set aside forest land from which products can be obtained for sale in local markets to obtain monetary income. The programs reserved for areas of Conversion or Unclassified forest land do not consider options of forest exploitation, although they are required to dedicate some land to other uses.

For instance, the HTI estates have to dedicate 10% of the land to *tanaman kehiduppi,* which can be private vegetable or mixed fruit gardens. Similarly, PIR projects provide land to participating farmers for their own house gardens. Considering the objectives of estate development, alternative choices will have to provide the same incomes as the estate options to be socially acceptable. They have to he generated at least partly from some kind of exploitation of forests in order to be able to choose alternative approaches for developing swidden agriculture without losing its intrinsic biodiversity.

Diverse tree gardens can provide significant *levels* of income, demonstrated by the way rubber is still produced in species-rich gardens, typical for most smallholder rubber producers. In many isolated villages in Kalimantan, rubber remains the most important cash crop and low harvest levels may be as much the result of low yields, insufficient trees, or simply a limited need for cash. This is riot to say that rubber production should not be improved.

However, rather than trying to replace all traditional smallholder rubber gardens with improved rubber, or with nuclear estate rubber, replacement with clones that provide higher yields but which can be grown in traditional diverse rubber gardens, or in complex, tembawang-like forest gardens, should be considered also. Besides improving rubber yields from diverse species-rich rubber gardens, many other options are available to develop swidden agriculture without having to opt for monoculture production.

Reports exist from *Bidavuh* farmers in central West Kalimantan who gain substantial income from *durian* yields from *tembawang* forests. These fruits are being bought by traders from Kutching, Sarawak who travel for 6 h to reach the producers during the fruiting season.

Expanding markets in urban Indonesia, Malaysia and Singapore only will increase the demand for such products. Hence, the opportunity to increase incomes by producing durians in diverse forest gardens only will increase. Existing forest management practices, furthermore, provide a unique opportunity for smallholder timber production which eventually could sustain the supply of the local timber industry. Such timber can be planted in and harvested from rubber gardens, *tembawang, pulau rimba* and *hutan tutupan.*

The 33% of area under forest in the village where Dayak forest management was studied suggests the potential that such traditional forest management holds for the production of high quality timber. Some kind of tree cover creates a better environment to plant species which are naturally found in closed forests, rather than in open terrain where many seedlings die, or where fires may destroy plantations.

Additionally, in diverse holdings, other faster-growing species could be produced on small lots which would be rotated, for instance, with rice production in order to supply an industry that needs faster-growing species of lower wood quality. Monetary returns to small holders from developed swidden agricultural systems can be competitive with any of the other options pursued so far.

10 The Convention on Biological Diversity

The Contracting Parties

Conscious of the intrinsic value of biological diversity and of the ecological, genetic, social, economic, scientific, educational, cultural, recreational and aesthetic values of biological diversity and its components,

Conscious also of the importance of biological diversity for evolution and for maintaining life sustaining systems of the biosphere.

Affirming that the conservation of biological diversity is a common concern of humankind

Reaffirming that states have sovereign rights over their own biological resources

Reaffirming also that states are responsible for conserving their biological diversity and for using their biological resources in a sustainable manner,

Concerned that biological diversity is being significantly reduced by certain human activities,

Aware of the general lack of information and knowledge regarding biological diversity and of the urgent need to develop scientific, technical and institutional capacities to provide the basic understanding upon which to plan and implement appropriate measures,

Noting that it is vital to anticipate, prevent and attack the causes of significant reduction or loss of biological diversity at source,

Noting also that whether there is a threat significant reduction or loss of biological diversity, lack of full scientific certainty should not be used as a reason for postponing measures to avoid or minimise such a threat.

Noting further that the fundamental requirement for the conservation of biological diversity is the *in situ*, conservation of ecosystems and natural habitats and the maintenance and recovery of viable populations of species in their natural surroundings.

Noting further that *ex situ* measures, preferably in the country of origin, also have an important role to play.

Recognising the close and traditional dependence of any indigenous and local communities embodying traditional lifestyles on biological resources, and the desirability of sharing equitably benefits arising from the use of traditional knowledge, innovations and practices relevant to the conservation of biological diversity and the sustainable use of its components,

Recognising also the vital role that women play in the conservation and sustainable use of biological diversity and affirming the need for the full participation of women at all levels of policy-making and implementation for biological diversity conservation,

Stressing the importance of, and the need to promote, international, regional and global cooperation among States and intergovernmental organisations and the non governmental sector for the conservation of biological diversity and the sustainable use of its components,

Acknowledging further that special provision of new and additional financial resources and appropriate access to relevant technologies can be expected to make a substantial difference in the world's ability to address the loss of biological diversity,

Acknowledging further that special provision is required to meet the needs of developing countries, including the provision of new and additional financial resources and appropriate access to relevant technologies,

Noting in this regard the special conditions of the least developed countries and small island States,

Acknowledging that substantial investments are required to conserve biological diversity and that there is the expectation of a broad range of environmental, economic and social benefits from those investments,

Recognising that economic and social development and poverty eradication are the first and overriding priorities of developing countries,

Aware that conservation and sustainable use of biological diversity is of critical importance for meeting the food, health and other needs of the growing world population, for which purpose access to and sharing of both genetic resources and technologies are essential,

Noting that, ultimately, the conservation and sustainable use of biological diversity will strengthen friendly relations among States and contribute to peace for humankind,

Desiring to enhance and complement existing international arrangement for the conservation of biological diversity and sustainable use of its components, and

Determined to conserve and sustainable use biological diversity for the benefit of present and future generations,

Have agreed as follows:

Article 1. Objectives

The objectives of this Convention, to be pursued in accordance with its relevant provisions, are the conservation of biological diversity, the sustainable use of its components and the fair and equitable sharing of the benefits arising out of the utilisation of genetic resources, including by appropriate access to genetic resources and by appropriate transfer of relevant technologies, taking into account all rights over those resources and to technologies, and by appropriate funding.

Article 2. Use of Terms

For the purposes of this Convention:

"Biological diversity" means the variability among living organisms from all sources, including, inter alia, terrestrial, marine and other aquatic ecosystems and the ecological complexes of which they are part; this includes diversity within, species, between species and of ecosystems.

"Biological resources" includes genetic resources, organisms or parts thereof, populations, or any other biotic component of ecosystems with actual or potential use or value for humanity.

"Biotechnology" means any technological application that uses biological systems, living organisms, so derivatives therefore , to make or modify products or process for specific use.

"Country of origin of genetic resources" means the country which possesses those genetic resources is in situ conditions.

"Country providing genetic resources" means the country supplying genetic resources collected from in situ sources, including populations of both wild and domesticated species, or taken from ex situ sources, which may or may not have originated in that country "Domesticated or cultivated species" means species in which the evolutionary process has been influenced by humans to meet their needs.

"Ecosystem" means a dynamic complex of plant, animal and microorganism communities and their nonliving environment interacting as a functional unit.

"*Ex situ*, conservation" means the conservation of components of biological diversity outside their natural habitats.

"Genetic Material" means any material of plant, animal, microbial or other origin containing functional units of heredity.

"Genetic resources" means genetic material of actual or potential value.

"Habitat" means the place or type of site where an organism or population naturally occurs.

"*In situ* conditions" means conditions where genetic resources exist within ecosystems and natural habitats and the maintenance and recovery of viable populations of species ins their natural; surrounding and, in the case, of domesticated or cultivated species, in the surrounding where they have developed their distinctive properties.

"Protected area" means a geographically defined area which is designated or regulated and managed to achieve specific conservation objectives.

"Regional economic integration organisation" means an organisation constituted by sovereign States of a given region, to which its member States have transferred competence in respect of matters governed by this Convention and which has been duly authorised, an accordance with its internal procedures, to sign, ratify, accept, approve or accede to it.

"Sustainable use" The use of components of biological diversity in a way and at a rate that does not lead to long term decline of biological diversity, thereby maintaining its potential to meet the needs and aspirations of present and future generations.

"Technology" includes biotechnology.

Article 3. Principle

States have, in accordance with the Charter of the United Nations and the principles of international law, the sovereign right to exploit their own resources pursuant to their own environmental policies, and the responsibility to ensure that activities within their jurisdiction or control for not cause damage to the environment of other States or of areas beyond the limits of national Jurisdiction.

Article 4. Jurisdictional Scope

Subject to the rights of other States, and except as otherwise expressly provided in this Convention, the provisions of this Convention apply, in relation to each Contracting Party:

a. In the case of components of biological diversity, in areas within the limits of its national jurisdiction; and

b. In the case of processes and activities, regardless of where their effects occur, carried out under its jurisdiction or control, within the area of its national jurisdiction or beyond the limits of national jurisdiction.

Article 5. Cooperation

Each Contracting Party shall, as far as possible and as

appropriate, cooperate with other contracting parties, directly or, where appropriate, through competent international organisations in respect of areas beyond national jurisdiction and on other matters of mutual interest, for the conservation and sustainable use of biological diversity.

Article 6. General Measures for Conservation and Sustainable Use

Each Contracting Party shall, in accordance with its particular conditions and capabilities:

a. Develop national strategies, plans or programmes for the conservation and sustainable use of biological diversity or adapt for this purpose existing strategies, plans or programmes which shall reflect, inter alia, the measures set out in this Convention relevant to the Contracting Party concerned; and

b. Integrate as far as possible and as appropriate, the conservation and sustainable use of biological diversity into relevant sectoral or cross-sectoral plans, programmes and policies.

Article 7. Identification and Monitoring

Each Contracting Party shall, as far as possible and as appropriate, in particular for the purposes of Articles 8 to 10:

a. Identify components of biological diversity important for its conservation and sustainable use having regard to the indicative list of categories set down in Annex I:

b. Monitor, through sampling and other techniques, the components of biological diversity identified pursuant to subparagraph (a) above, paying particular attention to those requiring urgent conservation measures and those which offer the greatest potential for sustainable use;

Identify process and categories of activities which have or are likely to have significant adverse impacts on the conservation and sustainable use of biological diversity, and monitor their effects through sampling and other techniques; and

Maintain and organise by any mechanism data derived from identification and monitoring activities pursuant to subparagraphs (a), (b) and (c) above.

Article 8. *In situ* Conservation

Each Contracting Party shall, as far as possible and as appropriate:

a. Establish a system of protected areas or areas where special measures need to be taken to conserve biological diversity;

b. Develop, where necessary, guidelines for the selection, establishment and management of protected areas where special measures need to be taken to conserve biological diversity;

c. Regulate or manage biological resources for the conservation of biological diversity within or outside protected areas with a view to ensuring their conservation and sustainable use;

d. Promote the protection of ecosystems, natural habitats and maintenance of viable populations of species in natural surroundings;

e. Promote environmentally sound and sustainable development in areas adjacent to protected areas with a view to furthering protection of these areas;

f. Rehabilitate and restore degraded ecosystems and promote the recovery of threatened species, inter alia, through the development and implementation of plans or other management strategies;

g. Establish or maintain means to regulate, manage or control the risks associated with the use and release of living modified organisms resulting from biotechnology which are likely to have adverse environmental impacts that could affect the conservation and sustainable use of biological diversity, taking also into account the risks to human health;

h. Prevent the introduction of, control or eradicate those alien species which threaten ecosystems, habitats or species;

i. Endeavour to provide the conditions needed or camptibility between present uses and the conservation of biological diversity and the sustainable use of its components

j. Subject to its national legislation, respect, preserve and maintain knowledge, innovations and practices of indigenous and local communities embodying traditional lifestyles relevant for the conservation and ustinable use of biological diversity and promote their wider application with the approval and involvement of the holders of such knowledge, innovations and practices and encourage the equitable sharing of the benefits arising from the utilisation of such knowledge, innovations and practices;

k. Develop or maintain necessary legislation and/or other regulatory provisions for the protection of threatened species and populations;

l. Where a significant adverse effect on biological diversity has been determined pursuant to Article 7, regulate or manage the relevant processes and categories of activities; and

m. Cooperate in providing financial and other support for in situ conservation outlined in subparagraphs (a) to (1) above, particularly to developing countries.

Article 9. Ex Situ Conservation

Each Contracting Party shall, as far as possible and as appropriate, and predominantly for the purpose of complementing in situ measures.

a. Components of biological diversity, of such components;

b. Conservation of and research on plants, animals and microorganisms;

c. Adopt measures for the recovery and rehabilitation of threatened species and for their reintroduction into appropriate conditions;

d. Regulate and manage collection of biological resources from purposes so as not to threaten ecosystem and in situ population of species except where special temporary ex situ measures are required under subparagraph (c) above; and

e. Cooperate in providing for ex situ conservation outlined in subparagraphs (a) to (d) above and in the establishment and maintenance of ex situ conservation facilities in developing countries.

Article 10 Sustainable use of components of biological diversity

Each Contracting Party shall, as far as possible and as appropriate

a. Integrate Consideration of the conservation and sustainable use of biological resources;

b. Adopt measures relating to the use of biological resources to avoid or minimise adverse impacts on biological diversity;

c. Protect and encourage customary use of biological resources to avoid or minimise adverse impacts on biological diversity;

d. Support local populations to develop and implement remedial action in degraded areas where biological diversity has been reduced; and

e. Encourage cooperation between its governmental authorities and its private sector in developing methods for sustainable use of biological resources.

Article 11. Incentive Measures

Each Contracting Party shall, as far as possible and as appropriate, adopt economically and socially sound measures that act as incentives for the conservation and sustainable use of components of biological diversity.

Article 12. Research and Training

The Contracting Parties, taking into account the special needs of developing countries, shall:

a. Establish and maintain programmes for scientific and technical education and training in measures for the identification, conservation and sustainable use of biological diversity and its components and provide support for such education and training for the specific needs of developing countries

b. Promote and encourage research which contributes to the conservation and sustainable use of biological diversity, particularly in developing countries inter alia, in accordance with decisions of the Conference of' the Parties taken consequence of recommendations of the subsidiary Body on Scientific, Technical and Advice; and

c. In keeping with the provisions of Articles 10, 18 and 20, promote and cooperate in the use of scientific advances in biological diversity research in developing methods for conservation and sustainable use of biological resources.

Article 13. Public Education and Awareness

The Contracting Parties shall:

a. Promote and encourage understanding of the importance of, and the measures required for, the conservation of biological diversity, as well as its propagation through media, and the inclusion of these topics in educational programmes; and

b. Cooperate, as appropriate, with other States and international organisations in developing educational and public awareness programmes, with respect to conservation and sustainable use of biological diversity.

Article 14. Impact Assessment and Minimising Adverse Impacts

I. Each Contracting Party, as far as possible and as appropriate, shall:

a. Introduce appropriate procedures requiring environmental impact assessment of its proposed projects that are likely to have significant adverse effects on biological diversity with a view to avoiding or minimising such effects and, where appropriate, allow for public participation in such procedures.

b. Introduce appropriate arrangements to ensure that the environmental consequence, of its programmes and policies that are likely to have significant adverse impacts on biological diversity are duly taken into account;

c. Promote, on the basis of reciprocity, notification, exchange of information and consultation on activities under their jurisdiction or control which are likely significantly to affect adversely the biological diversity of other States or areas beyond the limits of national jurisdiction, by encouraging the conclusion of bilateral, regional or multilateral arrangements as appropriate;

d. In the case of imminent or grave danger or damage originating under its jurisdiction or control to biological diversity within the area under jurisdiction of other States or in areas beyond the limits of national jurisdiction of other States or in areas beyond the limits of national jurisdiction, notify immediately the potentially affected States of such danger or damage, as well as initiate action to prevent or minimise such danger or damage;

e. Promote national arrangements for emergency responses to activities or events, where caused naturally or otherwise, which present a grave and imminent danger to biological diversity and encourage international cooperation to supplement such national efforts and, where appropriate and agreed by the States or regional economic integration organisations concerned, to establish joint contingency plans.

2. The Conference of the Parties shall examine, on the basis of studies to be carried out, the issue of liability and redress, including restoration and compensation, for damage to biological diversity, except where such liability is a purely internal matter.

Article 15. Access to Genetic Resources

1. Recognising the sovereign rights of States over their natural resources, the authority to determine access to genetic resources rests with the national governments and is subject to national legislation.

2. Each Contracting Party shall endeavour to create conditions to facilitate access to genetic resources for environmentally sound uses by other Contracting Parties and to impose restrictions that run counter to the objectives of this Convention.
3. For the purpose of this Convention, the genetic resources being provided by a Contracting Party, as referred to in this Article and Article 16 and 19, origin of such resources or by the Parties that have acquired the genetic resources in complete with this Convention.
4. Access, where granted, shall be on mutually agreed terms and subject to the provisions of the Article.
5. Access to genetic resources shall be subject to prior informed consent of the Contracting Party providing such resources, unless otherwise determined by that Party.
6. Each Contracting Party shall endeavour to develop and carry out scientific research based on genetic resources provided by other Contracting Parties with the full participation of, and where possible in, such Contracting Parties.
7. Each Contracting Party shall take legislative, administrative or policy measures, as appropriate, and in accordance with Articles 16 and 19 and where necessary through the financial mechanism established by Articles 20 ant 21, with the aim of starting in a fair and equitable way the results of research and development and the benefits arising from the commercial and other utilisation of genetic resources with the Contracting Party providing such resources. Such sharing shall be upon mutually agreed terms.

Article 16. Access to and Transfer of Technology

1. Each Contracting Party, recognising that technology includes biotechnology, and that both access to and transfer of technology among Contracting Parties are essential elements for the attainment of the objectives of this Convention, undertakes subjects to the provisions of this Article to provide and/or facilitate access for an transfer to other Contracting Parties of technologies that are relevant to the conservation and sustainable use of biological diversity or make use of genetic resources and do not cause significant damage to the environment.

2. Access to and transfer of technology referred to in paragraph 1 above to developing countries shall be provided and/or facilitated under fair and most favourable terms, including on concessional and preferential terms where agreed, and where necessary in accordance with the financial mechanism established by Articles 20 and 21. In the case of technology to patents and other intellectual property rights, such access and transfer shall be provided on terms with recognise and are consistent with the adequate and effective protection of intellectual property rights. The application of this paragraph shall he consistent with paragraphs 3,4 and 5 below.

3. Each Contracting Party shall take legislative, administrative or policy measures, as appropriate, with the aim that Contracting Parties, in particular those that are developing countries, which provide genetic resources are provided access to and transfer of technology which makes use of those resources,, on mutually agreed terms, including technology protected by patents and other intellectual property rights, where necessary through the provisions of

Articles 20 and 21 and in accordance with international law and consistent with paragraphs 4 and 5 below.

4. Each Contracting Party shall take legislative, administrative or policy measures, as appropriate, with the aim that the private sector facilitates access to, joint development and transfer of technology referred to in paragraph 1 above for the benefit of both governmental institutions and the private sector of developing countries and in this regard shall abide by the obligations included in paragraphs 1, 2, and 3 above.

5. The Contracting Parties, recognising that patents and other intellectual property rights may have an influence on the implementation of this Convention, shall cooperate in this regard subject to national legislation and international law in order to ensure that such rights are supportive of and do not run counter to its objectives.

Article 17. Exchange of Information

1. The Contracting Parties shall facilitate the exchange of information from all publicly available source, relevant to the conservation and sustainable use of biological diversity, taking into account the special needs of developing countries.

2. Such exchange of information shall include exchange of results of technical, scientific and socioeconomic research, as well as information on training and surveying programmes, specialised knowledged, indigenous and traditional knowledge as such and in combination with the technologies referred to in Article 16, paragraph 1. It shall also, where feasible, include repatriation of information.

Article 18. Technical and Scientific Cooperation

1. The Contracting Parties shall promote international technical and scientific cooperation in the field of conservation and sustainable use of biological diversity, where necessary through the appropriate international and national institutions.

2. Each Contracting Party shall promote technical in particular developing countries, in implementing this Convention, inter alia, through the development and implementation of national policies. In promoting such cooperation, special attention should be given to the development and strengthening of national capabilities by means of human resources development and institution building.

3. The Conference of the Parties, at its first meeting, shall determine how to establish a clearinghouse mechanism to promote and facilitate technical and scientific cooperation.

4. The Contracting Parties shall, in accordance with national legislation and policies, encourage and develop methods of cooperation for the development and use of technologies, including indigenous and traditional technologies, in pursuance of the objectives of this Convention. For this purpose, the Contracting Parties shall also promote cooperation in the training of personnel and exchange of experts.

5. The Contracting Parties shall, subject to mutual agreement, promote the establishment of joint research programmes and joint ventures for the development of technologies relevant to the objectives of this Convention.

Article 19. Handling of Biotechnology and Distribution of its Benefits

1. Each Contracting Party shall take legislative, administrative or policy measures, as appropriate, to provide for the effective participation in biotechnological research activities by those Contracting Parties, especially developing countries, which provide the genetic resources for such research, and where feasible in such Contracting Parties.

2. Each Contracting Party shall take all practicable measures to promote and advance priority access on a fair and equitable basis by Contracting Parties, especially developing countries, to the results and benefits arising from biotechnologies based upon genetic resources provided by those Contracting Parties. Such access shall be on mutually agreed terms.

3. The Parties shall consider the need for and modalities of a protocol setting out appropriate procedures, including, in particular, advance informed agreement, in the field of the safe transfer, handling and use of any living modified organism resulting from biotechnology that may have adverse effect on the conservation and sustainable use of biological diversity.

4. Each Contracting Party shall, directly or by requiring any natural or legal person under its jurisdiction providing the organisms referred to in paragraph 3 above, provide any available information about the use and safety regulations required by that Contracting Party in handling such organisms as well as any available information on the potential adverse impact of the specific organisms concerned to the Contracting Party into which those organisms are to be introduced.

Article 20. Financial Resources

1. Each Contracting Party undertakes to provide, in accordance with its capabilities, financial support and incentives in respect of those national activities which are intended to achieve the objectives of this Convention, in accordance with its national plans, priorities and programmes.

2. The developed country Parties shall provide new and additional financial resources to enable developing country Parties to meet the agreed full incremental costs to them of implementing measures which fulfil the obligations of this Convention and to benefit from Its provisions and which costs are agreed between a developing country Party and the institutional structure, referred to in Article 21, in accordance with policy, strategy, programme priorities and eligibility criteria and an indicate list of incremental costs established by the

   Conference of the Parties. Other Parties,, including countries undergoing the process of transition to a market economy, may voluntarily assume the obligations of the developed country Parties. For the purpose of this Article, the Conference of' the Parties shall at its first meeting establish a list of developed country Parties and other Parties which voluntarily assume the obligations of the developed country Parties. The Conference of the Parties shall periodically review and if necessary amend the list. Contributions from other countries and source on a voluntary basis would also be encouraged. The implementation of these commitments shall take into account the need for adequacy, predictability and timely flow of funds and the importance of burden-sharing among the contributing Parties included in the list.

3. The developed country Parties may also provide, and developing country Parties avail themselves of, financial resources related to the implementation of this Convention through bilateral, regional and other multilateral channels.

4. The extent to which developing country Parties will effectively implement their commitments under the Convention will depend on the effective implementation by developed country Parties of their commitments sunder the Convention related to financial resources and transfer of technology and will take fully into account the fact that economic and social development and eradication of poverty are the first and overriding priorities of the developing country Parties.

5. The Parties shall take full account of the specific needs and special situation of least developed countries in their actions with regard to funding and transfer of technology.

6. The Contracting Parties shall also take into consideration the special conditions resulting from the dependence on distribution and location of biological diversity within developing country Parties, in particular small island States.

7. Consideration shall also be given to the special situation of developing countries, including those that are most environmentally vulnerable, such as those with arid and semi-arid zones, coastal and mountainous areas.

Article 21. Financial Mechanism

1. There shall be a mechanism for the provision of financial resources to developing country Parties for purposes of this Convention on a grant or

concessional basis the essential elements of which are described in this Article. The mechanism shall function under the authority and guidance of, and be accountable to, the Conference of the Parties for purposes of this Convention. The operations of mechanism shall be carried out by such institutional structure as may be decided upon by the Conference of the Parties at its first meeting. For purposes of this Convention, the Conference of the Parties shall determine the policy, strategy, programme priorities and eligibility criteria relating to the access to and utilisation of such resources. The contributions shall be such as to take into account the need for predictability, adequacy and timely flow of funds referred to in Article 20 in accordance with the amount of resources needed to be decided periodically by the Conference of the Parties included in the list referred to in Article 20, paragraph 2. Voluntary contributions may also be made by the developed country parties and by other countries and sources. The mechanism shall operate within a democratic and transparent system of governance.

2. Pursuant to the objectives of this Convention, the Conference of the Parties shall at its first meeting determine the policy, strategy and programme priorities, as well as detailed criteria and guidelines for eligibility for access to and utilisation of the financial resources including monitoring and evaluation on a regular of such utilisation. The Conference of the Parties shall decide on the arrangements to give effect to paragraph 1 above after consultation with the institutional structure entrusted with the operation of the financial mechanism.

3. The Conference of the Parties shall review the effectiveness of the mechanism established under this Article, including the criteria and guidelines referred to in paragraph 2 above, not less than two years after the entry into force of this Convention and thereafter on a regular basis. Based on such review, it shall take appropriate action to improve the effectiveness of the mechanism if necessary.

4. The Contracting Parties shall consider strengthening existing financial institutions to provide financial resources for the conservation and sustainable use of biological diversity.

Article 22. Relationship with other International Conventions

1. The provisions of the present Convention shall not affect the rights and obligations of any Contracting Party deriving from any existing international agreement, except where the exercise of those rights and obligations would cause serious damage or a threat to biological diversity.

2. Contracting Parties shall implement this Convention with respect to the marine environment consistently with the rights and obligations of States under the law of the sea.

Article 23. Conference of the Parties

1. A Conference of the Parties is hereby established. The first meeting of the Conference of the Parties shall be convened by the Executive Director of the United Nations Environment Programme not later than one year after the entry into force of this Convention. Thereafter, ordinary meetings of the Conference of the Parties shall be held at regular intervals to be determined by the Conference at its first meeting.

2. Extraordinary meetings of the Conference of the Parties shall be held at such other times as may be deemed necessary by the Conference or at the written request of any Party, provided that within six months of the request being communicated to them by the Secretariat, it is supported by at least one-third of the Parties.

3. The Conference of the Parties shall by consensus agree upon and adopt rules of procedure for itself and for any subsidiary body it may establish, as well as financial rules governing the funding of the Secretariat. At each ordinary meeting, it shall adopt a budget for the financial period until the next ordinary meeting.

4. The Conference of the Parties shall keep under review the implementation of the present Convention, and, for this purpose, shall:

    a. Establish the form and the intervals for transmitting the information to be submitted in accordance with Article 26 and consider such information as well as reports submitted by any subsidiary body;

    b. Review scientific, technical and technological advice on biological diversity provided in accordance with Article 25;

    c. Consider and adopt, as required, protocols in accordance with Article 28;

    d. Consider and adopt, as required, in accordance with Articles 29 and 30, amendments to this Convention and its annexes;

    e. Consider amendments to any protocol, as well as to any annexes thereto, and, if so decided, recommend their adoption to the parties to the protocol concerned;

f. Consider and adopt, as required, in accordance with Article 30, additional annexes to this Convention;

g. Establish such subsidiary bodies, particularly to provide scientific and technical advice, as are deemed necessary for the implementation of this Convention;

h. Contact, through the Secretariat, the executive bodies of conventions dealing with matters covered by this Convention with a view to establishing appropriate forms of cooperation with them; and

i. Consider and undertake any additional action that may be required for the achievement of the purposes of this Convention in the light of experience gained in its operation.

5. The United Nations, its specialised agencies and the International Atomic Energy Agency, as well as any State not party to this Convention, may be represented as observers at meetings of the Conference of the Parties. Any other body or agency, whether governmental or nongovernmental, qualified in fields relating to conservation and sustainable use of biological diversity which has informed the Secretariat of its wish to be represented as an observer at a meeting of the Conference of the Parties present object.

Article 24. Secretariat

1. A secretariat is hereby established. Its functions shall be:

a. To arrange for and service meetings of the Conference of the Parties provided for in Article 23;

b. To perform the functions assigned to it by any protocol;

c. To prepare report on the execution of its function under this Convention and present them to the Conference of the Parties;

d. To coordinate with other relevant international bodies, and in particular to enter into such administrative and contractual arrangements as may be required for the effective discharge of its functions; and

e. To perform such other functions as may be determined by the Conference of the Parties.

2. At its first ordinary meeting, the Conference of the Parties shall designate the secretariat from amongst those existing competent international organisations which have signified their willingness to carry out the secretariat functions under this Convention.

Article 25. Subsidiary body on scientific, technical and technological advice

1. A subsidiary body for the provision of scientific, technical and technological advice is hereby established to provide the Conference of the Parties and, as appropriate, its other subsidiary bodies with timely advice relating to the implementation of this Convention. This body shall be open to participation by all Parties and shall be multi disciplinary. It shall comprise government representatives competent in the relevant field of expertise. It shall report regularly to the Conference of the Parties on all aspects of its work.

2. Under the authority of and in accordance with guidelines laid down by the Conference of the Parties, and upon its request, this body shall;

a. Provide scientific and technical assessments of the status of biological diversity:

b. Prepare scientific and technical assessments of the effects of types of measures taken in accordance with the provisions of this Convention

c. Identity innovative, efficient and state of art technologies and know-how relating to the conservation and sustainable use of biological diversity and advise on the ways and means of promoting development and/or transfering such technologies;

d. Provide advice on scientific programmes and international cooperation in research and development related to conservation and sustainable use of biological diversity; and

e. Respond to scientific, technical, technological and methodological questions that the Conference of the Parties and its subsidiary bodies may put to the body.

The functions terms of reference, organisation and operation of this body may be further elaborated by the Conference of the Parties.

Article 26. Reports

Each Contracting Party shall, at intervals to be determined by the Conference of the Parties, present to the Conference of the Parties reports on measures which it has taken for the implementation of the provisions of this Convention and their effectiveness in meeting the objectives of this Convention

Article 27. Settlement of Disputes

1. In the event of a dispute between Contracting Parties concerning the interpretation or application of this Convention, the Parties concerned shall seek solution by negotiation.

2. If the parties concerned cannot reach agreement by negotiation, they may jointly seek the good offices of, or request mediation by, a third party.

3. When ratifying, accepting, approving or acceding to this Convention, or at any time thereafter, a State or regional economic integration organisation may declare in writing to the Depositary that for a dispute not resolved in accordance with paragraph 1 or paragraph 2 above, it accepts one or both of the following means of dispute, settlement as compulsory:

   a. Arbitration in accordance with the procedure laid down in Part 1 of Annex 11;

   b. Submission of the dispute to the International Court of Justice.

4. If the parties to the dispute have not in accordance with paragraph 3 above, accepted the same or any procedure, the dispute shall be submitted to conciliation in accordance with Part 2 of Annex 11 unless the parties otherwise agree.

5. The provisions of this Article shall apply with respect to any protocol except as otherwise provided in the protocol concerned.

Article 28. Adoption of Protocols

1. The Contracting Parties shall cooperate in the formulation and adoption of protocols to this Convention.

2. Protocols shall be adopted at a meeting of the Conference of the Parties.

3. The text of any proposed protocol shall be communicated to the Contracting Parties by the Secretariat at least six months before such a meeting.

Article 29. Amendment of the Convention or protocols

1. Amendments to this Convention may be proposed by any Contracting Party. Amendments to any protocol may be proposed by any Party to that protocol.

2. Amendments to this Convention shall be adopted at a meeting of the Conference of the Parties. Amendments to any protocol shall be adopted at meeting of the Parties to the Protocol in question. The text of any proposed amendment to this Convention or to any protocol, except as may otherwise be provided in such protocol, shall be communicated to the parties to the instrument in question by the secretariat at least six months before the meeting at which it is proposed for adoption. The secretariat shall also communicate proposed amendments to the signatories to this Convention for information.

3 The Parties shall make every effort to reach agreement on any proposed amendment to this Convention or to any protocol by consensus. If all efforts at consensus have been exhausted, and no agreement reached, the amendment shall as a last resort be adopted by a two-thirds majority vote of the Parties to the instrument in question present and voting at the meeting, and shall be submitted by the Depositary to all Parties for ratification, acceptance or approval.

4. Ratification, acceptance or approval of amendments shall be notified to the Depositary in writing. Amendments adopted in accordance with paragraph 3 above shall enter into force among Parties having accepted them on the ninetieth day after the deposit of instruments of ratification, acceptance or approval by at least two-thirds of the Contracting Parties to

this Convention or of the Parties to the protocol concerned, except as may otherwise be provided in such protocol. Thereafter the amendments shall enter into force for any other Party on the ninetieth day after that Party deposits its instruments of ratification, acceptance or approval of the amendments.

5. For the purposes of this Article, "Parties present and voting" means Parties present and casting an affirmative or negative vote.

Article 30. Adoption and Amendment of Annexes

1. The annexes to this Convention or to any protocol shall form an integral part of this Convention or of such protocol, as the case may be, and, unless expressly provided otherwise, a reference to this Convention or its protocols constitutes at the same time a reference to any annexes thereto. Such annexes shall be restricted to procedural, scientific, technical and administrative matters.

2. Except as may be otherwise provided in any protocol with respect to its annexes, the following procedure shall apply to the proposal, adoption and entry into force of additional annexes to this Convention or of annexes to any protocol:

    a. Annexes to this Convention or to any protocol shall be proposed and adopted according to the procedure laid down in Article 29

    b. Any Party that is unable to approve an additional annex to this Convention or an annex to any protocol to which it is party shall so notify the Depositary, in writing, within one year from the date of the communication of the adoption by the Depositary. The Depositary shall without delay notify all Parties of any such notification received. A Party may at any time withdraw a previous declaration of objection and the annexes shall thereupon enter into force for that Party subject to subparagraph (c) below;

On the expiry of one year from the date of the communication of the adoption by the Depositary, the annex shall enter into force for all parties to this Convention or to any protocol concerned which have not submitted a notification in accordance with the provisions of subparagraph (b) above.

3. The proposal, adoption and entry into force of amendments to annexes to this Convention or to any protocol shall be subject to the same procedure as for the proposal, adoption and entry into force annexes to the Convention or annexes to any protocol.

4. If an additional annex or an amendment to an annex is related an amendment to this Convention or to any protocol, the additional annex or amendment shall not enter into force until such time as the amendment to this Convention or to the protocol concerned enters into force.

Article 31. Right to Vote

1. Except as provided for in paragraph 2 below, each Contracting Party to this Convention or to any protocol shall have one vote.

2. Regional economic integration organisations, in matters within them competence, shall exercise their right to vote if their member States which are Contracting Parties to the Convention of the relevant protocol. Such organisations shall not exercise their right to vote in their member States exercise theirs, and vice versa.

Article 32. Relationship between this Convention and its Protocols

1. A State or a regional economic integration organisation may not become a party to a protocol unless it is, or becomes at the same time a Contracting Party to this Convention.

2. Decisions under any protocol shall be taken only by the Parties to the protocol concerned. Any Contracting Party that has not ratified accepted or approved a protocol may participate as an observer in any meeting of the parties to that protocol.

Article 33. Signature

This Convention shall be open for signature at Rio de Janeiro by all States and any regional economic integration organisation from the fifth day of June 1992 until.

Article 34. Ratification, Acceptance or Approval

1. This Convention and any protocol shall be subject to ratification acceptance or approval by States and by regional economic integration organisations. Instruments of ratification, acceptance or approval shall be deposited with the Depository.

2. Any organisation referred to in paragraph 1 above which becomes a Contracting Party to this Convention or any protocol without any or its member States being a Contracting Party shall be bound by all the obligations under the Convention or the protocol, as the case may be. In the case of such organisations, one or more of whose member States is a Contracting Party to the Convention or relevant protocol the organisation and its member States shall decide on their respective responsibilities for the performance of their obligations under the Convention or protocol, as the case may be in such

cases, the organisation and the member States shall not be entitled to exercise rights under the Convention or relevant protocol concurrently.

3. In their instruments of ratification, acceptance or approval, the organisations referred to in paragraph 1 above shall declare the extent of their competence with respect to the matters governed by the Convention or the relevant protocol. These organisations shall also inform the Depositary of any relevant modification in the extent of their competence.

Article 35. Accession

1. This Convention and any protocol shall be open for accession by States and by regional economic integration organisations from the date on which the Convention or the protocol concerned is closed for signature. The instruments of accessions shall be deposited with the Depositary.

2. In their instruments of accession, the organisations referred to in paragraph 1 shall declare the extent of their competence with respect to the matters governed by the Convention or the relevant protocol. These organisations shall also inform the Depositary of any relevant modification in the extent of their competence.

3. The provisions of Article 34, paragraph 2, shall apply to regional economic integration organisations which accede to this Convention or any protocol.

Article 36. Entry into Force

1. This Convention shall enter into force on the ninetieth day after the date of deposit of the thirtieth instrument of ratification, acceptance approval or accession.

2. Any protocol shall enter into force on the ninetieth day after the date of deposit of the number of instruments of ratification, acceptance, approval or accession, specified in that protocol, has been deposited.

3. For each Contracting Party which ratifies, accepts or approves this Convention or accedes thereto after the deposit of the thirtieth instrument of ratification, acceptance, approval or accession, it shall enter into force on the ninetieth day after the date of deposit by such Contracting Party of its instrument of ratification, acceptance, approval or accession.

4. Any protocol, except as otherwise provided in such protocol, shall enter into force for a Contracting Party that ratifies, accepts or approves that protocol or accedes thereto after its entry into force pursuant to paragraph 2 above, on the ninetieth day after the date on which that Contracting Party deposits its instrument of ratification, acceptance, approval or accession, or on the date on which the Convention enters into force for that Contracting Party, whichever shall be the later.

5. For the purposes of paragraphs 1 and 2 above, any instrument deposited by a regional economic integration organisation shall not be counted as additional to those deposited by member States of such organisation.

Article 37. Reservations

No reservations may be made to this Convention.

Article 38. Withdrawals

1. At any time after two years from the date on which the present Convention has entered into force for a Contracting Party, that Contracting Party may

withdraw from the Convention by giving written notification to the Depositary.

2. Any such withdrawal shall take place upon expiry of one Year after the date of its receipt by the Depositary, or on such later date as may be specified in the notification of the withdrawal.

3. Any Contracting Party which withdraws from this Convention shall be considered as also having withdraw from any protocol to which it is party.

Article 39. Financial Interim Arrangement

Provided that it has been fully restructured in accordance with the requirements of Article 21, the Global Environment Facility of the United Nations Development Programme, the United Nations Environment Programme and the International Bank for Reconstruction and Development shall be the institutional structure referred to in Article 21 on an interim basis, for the period between the entry into force of this Convention and the first meeting of the Conference of the Parties or until the Conference of the Parties decides which institutional structure will be designated in accordance with Article 21.

Article 40. Secretariat interim arrangements

The secretariat to be provided by the Executive Director of the United Nations Environment Programme shall be the secretariat referred to in Article 24, paragraph 2, on an interim basis for the period between the entry into force of this Convention and the first meeting of the Conference of the Parties.

Article 41. Depositary

The Secretary-General of the United Nations shall assume the functions of Depositary of this Convention and any protocols.

Article 42. Authentic Texts

1. The original of this Convention, of which the Arabic, Chinese, English, French, Russian and Spanish texts are equally authentic, shall be deposited with the Secretary-General of the United Nations.

2. In witness whereof the undersigned, being duly authorised to that effect, have signed this Convention.

3. Done at Rio de Janeiro on this fifth day of June, one thousand nine hundred and ninety-two.

Annex 1

Identification and Monitoring

1. Ecosystems and habits: containing high diversity, large numbers of endemic or threatened species, or wilderness; required by migratory species of social, economic, cultural or scientific importance; or which are representative, unique or associated with key evolutionary or other biological processes;

2. Species and communities which are threatened, wild relatives of domesticated or cultivated species; of medical, agricultural or other economic value; or social, scientific or cultural importance; or importance for research into the conservation and sustainable use of biological diversity, such as indicator species; and

3. Described genomes and genes of social, scientific or economic importance.

Annex II

Part 1: Arbitration

Article 1

The claimant party shall notify the secretariat that the parties are referring a dispute to arbitration pursuant to

Article 27. The notification shall state the subject-matter of arbitration and include, in particular, the articles of the Convention or the protocol, the interpretation or application of which are at issue. If the parties do not agree on the subject matter of the dispute before the President of the tribunal is designated, the arbitral tribunal shall determine the subject matter. The secretariat shall forward the information thus received to all Contracting Parties to this Convention or to the protocol concerned.

Article 2

1. In disputes between two parties, the arbitral tribunal shall consist of three members. Each of the parties to the dispute shall appoint an arbitrator and two arbitrators so appointed shall designate by common agreement the third arbitrator who shall be the President of the tribunal. The latter shall not be a national of one of the parties to the dispute, nor have his or her usual place of residence in the territory of one of these parties, nor be employed by any of them, nor have dealt with the case in any other capacity.

2. In disputes between more than two parties, parties in the same interest shall appoint one arbitrator jointly by agreement.

3. Any vacancy shall be filled in the manner prescribed for the initial appointment.

Article 3

1. If the President of the arbitral tribunal has not been designated within two months of the appointment of the second arbitrator, the Secretary-General of the United Nations shall, at the request of a party, designate the president within a further two month period.

2. If one of the parties to the dispute does not appoint an arbitrator within two months of receipt to the request, the other party may inform the Secretary-General, who shall make the designation within a further two month period.

Article 4

The arbitral tribunal shall render its decisions in accordance with the provisions of this Convention, any protocols concerned, and international law.

Article 5

Unless the parties to the dispute otherwise agree, the arbitral tribunal shall determine its own rules of' procedure.

Article 6

The arbitral may, at the request of one of the parties, recommend essential interim measures of protection.

Article 7

The parties to the dispute shall facilitate the work of the arbitral tribunal and, in particular, using all means at their disposal, shall:

a. Provide it with all relevant documents, information and facilities; and
b. Enable it, when necessary, to call witnesses or experts and receive their evidence.

Article 8

The parties and the arbitrators are under an obligation to protect the confidentiality of any information they receive in confidence during the proceedings of the arbitral tribunal. *Article 9* Unless the arbitral tribunal determines otherwise because of the particular circumstances of the case, the costs of the tribunal shall be borne by the parties to the dispute in equal shares. The tribunal shall keep a record of all its costs, and shall furnish a final statement thereof to the parties.

Article 10

Any Contracting Party that has an interest of a legal nature in the subject-matter of the dispute which may be affected by the decision in the case may intervene in the proceedings with the consent of the tribunal.

Article 11

The tribunal may hear and determine counterclaims arising directly out of the subject-matter of the dispute.

Article 12

Decisions both on procedure and substance of the arbitral tribunal shall be taken by a majority vote of its members.

Article 13

If one of the parties to the dispute does not appear before the arbitral tribunal or fails to defend its case, the other party may request the tribunal to continue the proceedings and to make its award. Absence of a party or a failure of a party to defend its case shall not constitute a bar to the proceeding. Before rendering its final decision, the arbitral tribunal must satisfy itself that the claim is well founded in fact and law.

Article 14

The tribunal shall render its final decision within five months of the date on which it is fully constituted, unless it finds it necessary to extend the time-limit for a period which should not exceed five more months.

Article 15

The final decision of the arbitral tribunal shall be confined to the subject-matter of the dispute and shall state the reasons on which it is based. It shall contain the names of the members who have participated and the date of the final decision. Any member of the tribunal may attach a separate or dissenting opinion to the final decision.

Article 16

The award shall be binding on the parties to the dispute. It shall be without appeal unless the parties to the dispute have agreed in advance to an appellate procedure.

Article 17

Any controversy which may arise between the parties to the dispute as regards the interpretation or manner of implementation of the final decision may be submitted by either party for decision to the arbitral tribunal which rendered it.

Part 2: Conciliation

Article 1

A conciliation commission shall be created upon the request of one of the parties to the dispute. The commission shall, unless the parties otherwise agree, be composed of five members, two appointed by each Party concerned and a President chosen jointly by those members.

Article 2

In disputes between more than two parties, parties in the same interest shall appoint their members of the commission jointly by agreement. Where two or more parties have separate interests or there is a disagreement as to whether they are of the same interest, they shall appoint their members separately.

Article 3

If any appointments by the parties are not made within two months of the date of the request to create a conciliation commission, the Secretary-General of the United Nations shall, if asked to do so by the party that made the request, make those appointments within a further two month period.

Article 4

If a President of the conciliation commission has not been chosen within two months of the last of the members of the commission being appointed, the Secretary-General of the United Nations shall, if asked to do so by a party, designate a president within a further two month period.

Article 5

The conciliation commission shall take its decisions by majority vote of its members. It shall, unless the parties to the disputes otherwise agree, determine its own procedure. It shall render a proposal for resolution of the dispute, which the parties shall consider in good faith.

Article 6

A disagreement as to whether the conciliation commission has competence shall be decided by the commission.

11 ASEAN Agreement on the Conservation of Nature and Natural Resources (1985)

The Government of Negara Brunei Darussalam, The Government of the Republic of Indonesia, The Government of Malaysia, The Government of the Republic of Philippines, The Government of the Republic of the Singapore and The Government of the Kingdom of Thailand, Member States of the Association of South East Asian Nations (ASEAN).

Recognizing the importance of natural resources for present and future generations;

Conscious of their ever-growing value from a scientific, cultural, social and economic point of view;

Conscious also that the interrelationship between conservation and socio-economic development implies both that conservation is necessary to ensure sustainability of development, and that socio-economic development is necessary for the achievement of conservation on a lasting basis;

Recognizing the interdependence of living resources, between them and with other natural resources, within ecosystems of which they are part;

Wishing to undertake individual and joint action for the conservation and management of their living resources and the other natural elements on which they depend;

Recognizing that international co-operation is essential to attain many of these goals;

Convinced that an essential means to achieve such concerted action is the conclusion and implementation of an Agreement;

Have agreed as follows:

Chapter I

Conservation and Development

Article 1

Fundamental Principle

1) The Contracting Parties, within the framework of their respective national laws, undertake to adopt singly, or where necessary and appropriate through concerted action, the measures necessary to maintain essential ecological processes and life-support systems, to preserve genetic diversity, and to ensure the sustainable utilization of harvested natural resources under their jurisdiction in accordance with scientific principles and with a view to attaining the goal of sustainable development.

2) To this end they shall develop national conservation strategies, and shall co-ordinate such strategies within the framework of a conservation strategy for the Region.

Article 2

Development Planning

1) The Contracting Parties shall take all necessary measures, within the framework of their respective

national law, to ensure that conservation and management of natural resources are treated as an integral part of development planning at all stages and at all levels.

2) To that effect they shall, in the formulation of all development plans, give as full consideration to ecological factors as to economic and social ones.

3) The Contracting Parties shall, where necessary, take appropriate action with a view to conserving and managing natural resources of significant importance for two or several Contracting Parties.

Chapter II

Conservation of Species and Ecosystems

Article 3

Species - genetic diversity

1) The Contracting Parties shall, wherever possible, maintain maximum genetic diversity by taking action aimed at ensuring the survival and promoting the conservation of all species under their jurisdiction and control.

2) To that end, they shall adopt appropriate measures to conserve animal and plant species whether terrestrial, marine and freshwater, and more specifically

   (a) conserve natural, terrestrial, freshwater and coastal or marine habitats;

   (b) ensure sustainable use of harvested species;

   (c) protect endangered species;

   (d) conserve endemic species; and

   (e) take all the measures in their power to prevent the extinction of any species or sub-species.

3) In order to fulfil the aims of the preceding paragraph of this Article the Contracting Parties shall, in particular, endeavour to

   (a) create and maintain protected areas;

   (b) regulate the taking of species and prohibit unselective taking methods;

   (c) regulate and, where necessary, prohibit the introduction of exotic species;

   (d) promote and establish gene banks and other documented collections of animal and plant genetic resources.

Article 4

Species - Sustainable use

The Contracting Parties shall pay special attention to harvested species, and, to that effect, shall endeavour to

1) develop, adopt and implement management plans for those species, based on scientific studies and aiming at

   (a) preventing decrease in the size of any harvested population to levels below those which ensure its stable recruitment and the stable recruitment of those species which are dependent upon, or related to them;

   (b) maintaining the ecological relationship between harvested, dependent and related populations of living resources of the ecosystem considered;

   (c) restoring depleted populations to at least the levels referred to in sub-paragraph (1) of this paragraph;

(d) preventing changes or minimizing risk of changes in the ecosystem considered which are not reversible over a reasonable period of time.

2) Take the appropriate and necessary legislative and administrative measures on harvesting activities in the light of their national interest whereby

(a) such activities must conform to the management plans referred to above;

(b) the conduct of such activities is controlled by a permit system;

(c) all indiscriminate means of taking and the use of all means capable of causing local extinction of, or serious disturbance to, populations of a species or related species are prohibited;

(d) such activities are prohibited or strictly regulated at certain periods, seasons or places of importance in the life cycle of the species;

(e) such activities may be regulated more strictly, temporarily or locally in order to assist restoration of population levels or counterbalance any threat caused by special circumstances;

(f) special measures, such as restocking, are provided for whenever the conservation status of a species so warrants;

(g) trade and possession of specimens or products of specimens are regulated whenever such regulations meaningfully contribute to the implementation of the harvesting regulations.

Article 5

Species - endangered and endemic

1) Appendix 1 to this Agreement shall list endangered

species recognized by the Contracting Parties as of prime importance to the Region and deserving special attention. The Appendix shall be adopted by a meeting of the Contracting Parties.

Accordingly, Contracting Parties shall, wherever possible.

(a) prohibit the taking of these species, except for exceptional circumstances by special allowance from the designated authorities of the Contracting Parties;

(b) regulate the trade in and possession of specimens and products of those species accordingly;

(c) especially protect habitat of those species by ensuring that sufficient portions are included in protected areas;

(d) take all other necessary measures to improve their conservation status, and restore their populations to the highest possible level.

2) Each Contracting Party shall, wherever possible, apply the above measures to species endangered at national level.

3) The Contracting Parties recognize their special responsibility in respect of species that are endemic to areas under their jurisdiction and shall undertake accordingly to take wherever possible, all the necessary measures to maintain the population of such species at the highest possible level.

Article 6

Vegetation Cover and Forest Resources

1) The Contracting Parties shall, in view of the role of vegetation and forest covers in the functioning of

natural ecosystems, take all necessary measures to ensure the conservation of the vegetation cover and in particular of the forest cover on lands under their jurisdiction.

2) They shall, in particular, endeavour to

(a) control clearance of vegetation; endeavour to prevent bush and forest fires; prevent overgrazing by, inter alia, limiting grazing activities to periods and intensities that will not prevent regeneration of the vegetation;

(b) regulate mining and mineral exploration operations with a view to minimizing disturbance of vegetation and to requiring the rehabilitation of vegetation after such operations;

(c) set aside areas as forest reserves, inter alia, with a view to conserve the natural forest genetic resources;

(d) in reforestation and afforestation planning avoid as far as possible monoculture causing ecological imbalance;

(e) designate areas whose primary function shall be the maintenance of soil quality in the catchment considered and the regulation of the quantity and quality of the water delivered from it;

(f) ensure, to the maximum extent possible the conservation of their natural forests, particularly mangroves with a view, inter alia, to maintaining maximum forest species diversity;

(g) develop their forestry management plans on the basis of ecological principles with a view to maintaining potential for optimum sustained yield and avoiding depletion of the resource capital.

Article 7

Soil

1) The Contracting Parties shall, in view of the role of soil in the functioning of natural ecosystems, take measures, wherever possible towards soil conservation, improvement and rehabilitation; they shall in particular, endeavour to take steps to prevent soil erosion and other forms of degradation, and promote measures which safeguard the processes of organic decomposition and thereby its continuing fertility.

2) To that effect, they shall, in particular, endeavour to

(a) establish land use policies aimed at avoiding losses, of vegetation cover, substantial soil losses, and damages to the structure of the soil;

(b) take all necessary measures to control erosion, especially as it may affect coastal or freshwater ecosystems lead to siltation of downstream areas such as lakes or vulnerable ecosystems such as coral reefs, or damage critical habitats, in particular that of endangered or endemic species;

(c) take appropriate measures to rehabilitate eroded or degraded soils including rehabilitation of soil affected by mineral exploitation.

Article 8
Water

1) The Contracting Parties shall, in view of the role of water in the functioning of natural ecosystems, take all appropriate measures towards the conservation of their underground the surface water resources.

2) They shall to that effect, in particular, endeavour to

   (a) undertake and promote the necessary hydrological research especially with a view to ascertaining the characteristics of each watershed;

   (b) regulate and control water utilization with a view of achieving sufficient and continuous supply of water for, inter alia, the maintenance of natural life supporting systems and aquatic fauna and flora;

   (c) when planning and carrying out water resource development projects take fully into account possible effects of such projects on natural processes or on other renewable natural resources and prevent or minimize such effects.

Article 9
Air

The Contracting Parties shall, in view of the role of air in the functioning of natural ecosystems, endeavour to take all appropriate measures towards air quality management compatible with sustainable development.

Chapter III

Conservation of Ecological Processes

Article 10

Environmental Degradation

The Contracting Parties, with a view to maintaining the proper functioning of ecological processes, undertake, wherever possible, to prevent, reduce and control degradation of the natural environment and , to this end, such endeavour to undertake, in addition to specific measures referred to in the following article;

(a) to promote environmentally sound agricultural practice by, inter alia, controlling the application of pesticides, fertilizers and other chemical products for agricultural development schemes, in particular for wetland drainage or forest clearance, pay due regard to the need to protect critical habitats as well as endangered and economically important species;

(b) to promote pollution control and the development of environmentally sound industrial processes and products;

(c) to promote adequate economic or fiscal incentives for the purposes of sub-paragraphs (a) and (b) above;

(d) as far as possible to consider the originator of the activity which may lead to environmental degradation responsible for its prevention, for rehabilitation and remedial measures required;

(e) to take into consideration, when authorizing activities likely to affect the natural environment, the foreseeable interactions between the new activities proposed and those already taking place in the same area, and the result of such interactions on the air, waters and soils of the area;

(f) to pay particular attention to the regulation of activities which may have adverse affects on processes which are ecologically essential or on areas which are particularly important or sensitive form an ecological point of view, such as the breeding and feeding grounds of harvested species.

Article 11

Pollution

The Contracting Parties, recognizing the adverse effect that polluting discharges or emissions may have on natural processes and the functioning of natural ecosystems as well as on each of the individual ecosystem components, especially animal and plants species, shall endeavour to prevent, reduce and control such discharges, emissions or applications, in particular by

(a) submitting activities likely to cause pollution of the air, soil, freshwater, or the marine environment, to controls which shall take into consideration both the cumulative effects of the pollutants concerned and the self-purificating aptitude of the recipient natural environment;

(b) making such controls conditional on, inter alia, appropriate treatment of polluting emissions; and

(c) establishing national environmental quality monitoring programmes, particular attention being paid to the effects of pollution on natural ecosystems, and co-operation in such programmes for the Region as a whole.

Chapter IV

Environmental Planning Measures

Article 12

Land Use Planning

1) The Contracting Parties shall, wherever possible in the implementation of their development planning,

give particular attention to the national allocation of land usage. They shall endeavour to take the necessary measures to ensure the integration of natural resources conservation into the land use planning process and shall, in the preparation and implementation of specific land use plans at all levels, give as full consideration as possible to ecological factors as to economic and social ones. In order to achieve optimum sustainable land use, they undertake to base their land use plans as far as possible on the ecological capacity of the land.

2) The Contracting Parties shall, in carrying out the provisions of paragraph (1) above, particularly consider the importance of retaining the natural high productivity of areas such as coastal zones and wetlands.

3) They shall, where appropriate, coordinate their land use planning with a view to conserving and managing natural resources of significant importance for two or several Contracting Parties.

Article 13

Protected Areas

1) The Contracting Parties shall as appropriate establish, in areas under their jurisdiction, terrestrial, freshwater, coastal or marine protection areas for the purpose of safeguarding

— the ecological and biological processes essential to the functioning of the ecosystems of the Region;

— representative samples of all types of ecosystem of the Region;

— satisfactory population levels for the largest possible number of species of fauna and flora belonging to those ecosystems;

— areas of particular importance because of their scientific, educational, aesthetic, or cultural interest;

and taking into account their importance in particular as;

— the natural habitat of species of fauna and flora, particularly rare or endangered or endemic species;

— zones necessary for the maintenance of exploitable stocks of economically important species;

— pools of genetic material and safe refuges for species, especially endangered ones;

— sites of ecological, aesthetic or cultural interest;

— reference sources of scientific research;

They shall, in particular, take all measures possible in their power to preserve those areas which are of an exceptional character and are peculiar to their country or the Region as well as those which constitute the critical habitats of endangered or rare species, of species that are endemic to a small areas and of species that migrate between countries of Contracting Parties.

2) Protected areas established pursuant to this Agreement shall be regulated and managed in such a way as to further the objectives of the purpose of which they have been created. Contracting Parties shall, wherever possible, prohibit within such protected areas activities which are inconsistent with such objectives.

3) Protected areas shall include

(a) *National Parks*

i) This expression denotes natural areas that are sufficiently large to allow for ecological self-regulation of one or several ecosystems, and which have not been substantially altered by human occupation or exploitation.

ii) National Parks shall be placed under public control, their boundaries shall not be altered nor shall any portion of any National Park be alienated except by the highest competent authority.

iii) National Parks shall be dedicated to conservation and to scientific, educational and recreational uses and the common welfare of the people.

(b) *Reserves*

(i) This expression denotes areas set aside for the purpose of preserving a specific ecosystem, the critical habitat of certain species of fauna or flora, a water catchment area or for any other specific purpose relating to the conservation of natural resources or objects or areas of scientific, aesthetic, cultural, educational or recreational interest.

ii) After reserves have been established their boundaries shall not be altered nor shall any portion of such reserves be altered nor shall any portion of such reserves be alienated except by the authority establishing them or by higher authority.

iii) Reserves shall be dedicated to the purposes for which they have been created and, in the light of the national interests of the

Contracting Parties, any activity inconsistent with such purposes shall be prohibited.

4) Contracting Parties shall, in respect of any protected areas established pursuant to this Agreement

(a) prepare a management plan and manage the area on the basis of this plan;

(b) establish, wherever appropriate, terrestrial or aquatic buffer zones that shall be located around protected areas and which, in the case of marine areas, may include coastal land areas or watersheds of rivers flowing into the protected area; in such buffer zones all activities that may have harmful consequences on the ecosystems that such areas purport to protect shall be prohibited or regulated and activities which are consistent with the purpose of the protected area shall be promoted.

5) Contracting Parties shall, in respect of any protected area established pursuant to this Agreement, endeavour to

(a) prohibit the introduction of exotic animal or plant species;

(b) prohibit the use or release of toxic substances or pollutants which could cause disturbance or damage to protected ecosystems or to the species they contain;

(c) to the maximum extent possible, prohibit or control any activity exercised outside protected areas when such an activity is likely to cause disturbance or damage to the ecosystems or species that such protected areas purport to protect.

6) Contracting Parties shall co-operate in the development of principles, objectives, criteria and guidelines for the selection, establishment and management of protected areas in the Region with a view to establishing a co-ordinated network of protected areas throughout the Region giving particular attention to those of regional importance. An Appendix containing such principles, objectives, criteria and guidelines shall be drawn up in the light of the best scientific evidence as adapted to the conservation requirements of the Region and shall be adopted by a meeting of Contracting Parties.

7) In addition to the establishment of the protected areas referred to in paragraph 3 of this Article, Contracting Parties shall promote, through the adoption of appropriate measures, the conservation of natural areas by private owners, community or local authorities.

Article 14

Impact Assessment

1) The Contracting Parties undertake that proposals for any activity which may significantly affect the natural environment shall as far as possible be subjected to an assessment of their consequences before they are adopted, and they shall take into consideration the results of this assessment in their decision-making process.

2) In those cases where any such activities are undertaken, the Contracting Parties shall plan and carry them out so as to overcome or minimize any assessed adverse effects and shall monitor such effects with a view to taking remedial action as appropriate.

Chapter V

National Supporting Measures

Article 15

Scientific Research

The Contracting Parties shall individually or in co-operation with other Contracting Parties or appropriate international organizations, promote and, whenever possible, support scientific and technical programmes of relevance to the conservation and management of natural resources, including monitoring, research, the exchange of technical information and the evaluation of results.

Article 16

Education, Information and Participation of the Public, Training

1) The Contracting Parties shall endeavour to promote adequate coverage of conservation and management of natural resources in education programmes at all levels.

2) They shall circulate as widely as possible information on the significance of conservation measures and their relationship with sustainable development objectives, and shall, as far as possible, organize participation of the public in the planning and implementation of conservation measures.

3) Contracting Parties shall endeavour to, individually or in co-operation with other Contracting Parties or appropriate international organizations, develop the programmes and facilities necessary to train adequate and sufficient scientific and technical personnel to fulfil the aims of this Agreement.

Article 17

Administrative Machinery

1) The Contracting Parties shall identify or maintain the administrative machinery necessary to implement the provisions of this Agreement, and, where several governmental institutions are involved, create the necessary co-ordinating mechanism for the authorities dealing with designated aspects of the environment.

2) They shall endeavour to allocate sufficient funds to the task necessary for the implementation of this Agreement, as well as sufficient qualified personnel with adequate enforcement powers.

Chapter VI

International Co-operation

Article 18

Co-operative Activities

1) The Contracting Parties shall co-operate together and with the competent international organizations, with a view to co-ordinating their activities in the field of conservation of nature and management of natural resources and assisting each other in fulfilling their obligations under this Agreement.

2) To that effect, they shall endeavour

(a) to collaborate in monitoring activities;

(b) to the greatest extent possible, co-ordinate their research activities;

(c) to use comparable or standardized research techniques and procedures with a view to obtaining comparable data;

(d) to exchange appropriate scientific and technical data, information and experience, on a regular basis;

(e) whenever appropriate, to consult and assist each other with regard to measures for the implementation of this Agreement.

3) In applying the principles of co-operation and co-ordination set for the above, the Contracting Parties shall forward to the Secretariat

(a) information of assistance in the monitoring of the biological status of the natural living resources of the Region;

(b) Information, including reports and publications of a scientific, administrative or legal nature and, in particular, information on

— measures taken by the Parties in pursuance of the provisions of this Agreement;

— the status of species included in Appendix l;

— any other matter to which the Conference of the Parties may give special priority.

Article 19

Shared Resources

1) Contracting Parties that share natural resources shall co-operate concerning their conservation and harmonious utilization, taking into account the sovereignty, rights and interests of the Contracting Parties concerned in accordance with generally accepted principles of international law.

2) To that end, they shall, in particular

(a) co-operate with a view to controlling, preventing, reducing or eliminating adverse environmental effects which may result in one Contracting Party from the utilization of such resources in another Party;

(b) endeavour to conclude bilateral or multilateral agreements in order to secure specific regulation of their conduct in respect of the resources concerned;

(c) as far as possible make environmental assessments prior to engaging in activities with respect of shared natural resources which may create a risk of significantly affecting the environment of another sharing Contracting Party or other sharing Contracting Parties;

(d) notify in advance the other sharing Contracting Party or the other sharing Contracting Parties of pertinent details of plans to initiate, or make a change in, the conservation or utilization of the resource which can reasonably be expected to affect significantly the environment in the territory of the other Contracting Party or Contracting Parties;

(e) upon request of the other sharing Contracting Party or sharing Contracting Parties, enter into consultation concerning the above mentioned Plans;

(f) inform the other sharing Contracting party or other sharing Contracting Parties of emergency situations or sudden grave natural events which may have repercussions on their environment;

(g) whenever appropriate, engage in joint scientific studies and assessments, with a view to facilitating co-operation with regard to environmental problems related to a shared resource, on the basis of agreed data.

3) Contracting Parties shall especially co-operate together and, where appropriate, shall endeavour to co-operate with other Contracting Parties, with a view to

(a) the conservation and management of

— border or contiguous protected areas;

— shared habitats of species listed in Appendix l;

— shared habitats of any other species of common concern;

(b) the conservation, management and, where applicable, regulation of the harvesting of species which constitute shared resources;

— by virtue of their migratory character, or

— because they inhabit shared habitats.

Article 20

Transfrontier Environmental Effects

1) Contracting Parties have in accordance with generally accepted principles of international law the responsibility of ensuring that activities under their jurisdiction or control do not cause damage to the environment of the natural resources under the jurisdiction of other Contracting Parties or of areas beyond the limits of national jurisdiction.

2) In order to fulfil this responsibility, Contracting Parties shall avoid to the maximum extent possible and reduce to the minimum extent possible adverse environmental effects of activities under their jurisdiction or control, including effects on natural resources, beyond the limits of their national jurisdiction.

3) To that effect, they shall endeavour

(a) to make environmental impact assessment before engaging in any activity that may create a risk of significantly affecting the environment or the natural resources of another Contracting Party or the environment or natural resources beyond national jurisdiction;

(b) to notify in advance the other Contracting Party or Contracting Parties concerned of pertinent details of plans to initiate, or make a change in, activities which can reasonably be expected to have significant effects beyond the limits of national jurisdiction;

(c) to enter into consultation concerning the above-mentioned plans upon request of the Contracting Party or Contracting Parties in question;

(d) to inform the Contracting Party or Contracting Parties in question of emergency situations or sudden grave natural events which may have repercussion beyond national jurisdiction.

4) contracting Parties shall, in particular, endeavour to refrain from actions which might directly or indirectly adversely affect wildlife habitats situated beyond the limits of national jurisdiction, especially habitats of species listed in Appendix 1 or habitats included in protected areas.

Chapter VII
International Supporting Measures

Article 21

Meeting of the Contracting Parties

1) Ordinary meetings of the Contracting Parties shall be held at least once in three years, in as far as possible in conjunction with appropriate meetings of ASEAN, and extraordinary meetings shall be held at any other time, upon the request of one Contracting Party provided that such request is supported by at least one other Party.

2) It shall be the function of the meetings of the Contracting Parties, in particular

(a) to keep under review the implementation of this Agreement and the need for other measures, in particular the Appendices;

(b) to adopt review and amend as required any Appendix to this Agreement;

(c) to consider reports submitted by the Contracting Parties in accordance with Article 28 or any other information which may be submitted by a Party, directly or through the Secretariat;

(d) to make recommendations regarding the adoption of any Protocol or any amendment to this Agreement;

(e) to establish working groups or any other subsidiary body as required to consider any matter related to this Agreement;

(f) to consider and undertake any additional action, including the adoption of financial rules, that may be required for the achievement of the purposes of this Agreement.

Article 22

Secretariat

On the coming into force of this Agreement the Contracting Parties shall designate the Secretariat responsible for carrying out the following functions:

(a) to convene and prepare the meetings of Contracting Parties;

(b) to convene diplomatic conferences for the purpose of adopting Protocols;

(c) to transmit to Contracting Parties notification, reports and other information received in accordance with this Agreement;

(d) to consider inquiries by, and information from, the Contracting Parties, and to consult with them on questions relating to this Agreement.

(e) to perform such other functions as may be assigned to it by the Contracting Parties;

(f) to ensure the necessary co-ordination with other competent international bodies and in particular to enter into such administrative arrangements as may be required for the effective discharge of the secretariat functions.

Article 23

National Focal Points

In order to facilitate communications with other Parties and the Secretariat, the Contracting Parties shall designate an appropriate national agency or institution responsible for co-ordinating matters arising from consultations and channeling communications between Contracting Parties or with the Secretariat.

Chapter VIII

Final Clauses

Article 24

Adoption of Protocols

1) The Contracting Parties shall co-operate in the formulation and adoption of Protocols to this Agreement, prescribing agreed measures, procedures and standards for the implementation of this Agreement.

2) The Contracting Parties, at a diplomatic conference, may adopt Protocols to this Agreement.

3) The Protocols of this Agreement shall be subject to acceptance and shall enter into force on the thirtieth day after the deposit with the Depositary of the Instruments of Acceptance of all the Contracting Parties.

Article 25

Amendment of the Agreement

1) Any Contracting Party to this Agreement may propose amendments to the Agreement. Amendments shall be adopted by a diplomatic conference which shall be convened at the request of the majority of the Contracting Parties.

2) Amendments to this Agreement shall be adopted by a consensus of the Contracting Parties.

3) Acceptance of amendments shall be notified to the Depositary in writing and shall enter into force on the thirtieth day following the receipt by the Depositary of notification of the acceptance by all the Contracting Parties.

4) After the entry into force of an amendment to this Agreement any new Contracting Party to this Agreement shall become a Contracting Party to this Agreement as amended.

Article 26

Appendices and Amendments to Appendices

1) Appendices to this Agreement shall form an integral part of the Agreement.

2) Amendments to an Appendix:

   (a) Any Contracting Party may propose amendments to an Appendix at a meeting of the Contracting Parties;

(b) Such amendments shall be adopted by a consensus of the Contracting Parties;

(c) The Depositary shall without delay communicate the amendment so adopted to all Contracting Parties.

3) The adoption and entry into force of a new Appendix to this Agreement shall be subject to the same procedure as for the adoption and entry into force of an amendment to an Appendix as provided for in paragraph (2) of this Article, Provided that the new Appendix shall not enter into force until such time as the amendment to the Agreement enters into force.

Article 27

Rules of Procedure

The Contracting Parties shall adopt rules of procedure for their meetings.

Article 28

Reports

The Contracting Parties shall transmit to the Secretariat reports on the measures adopted in implementation of this Agreement in such form and at such intervals as the meetings of Contracting Parties may determine.

Article 29

Relationships with Other Agreements

The provisions of this Agreement shall in no way affect the rights and obligations of any Contracting Party with regard to any existing treaty, convention or agreement.

Article 30

Settlement of Disputes

Any dispute between the Contracting Parties arising out

of the interpretation or implementation of this Agreement shall be settled amicably by consultation or negotiation.

Article 31

Ratification

This Agreement shall be subject to ratification by the Contracting Parties. The instruments of Ratification shall be deposited with the Secretary-General of the ASEAN Secretariat, who shall assume the functions of Depositary.

Article 32

Accession

1) After the entry into force of the Agreement, any Member State may accede to this Agreement, subject to prior approval by the Contracting Parties to this Agreement.

2) Instrument of accession shall be deposited with the Depositary.

Article 33

Entry into Force

1) This Agreement shall enter into force on the thirtieth day after the deposit of the sixth Instruments of Ratification.

2) Thereafter, this Agreement shall enter into force with respect to any Contracting Party on the thirtieth day following the date of deposit of the instrument of accession by that Contracting Party.

Article 34

Responsibility of the Depositary

The Depositary shall inform the Governments which have signed this Agreement;

(a) of the deposit of instruments of ratification, acceptance or accession;

(b) of the date on which the Agreement will come into force.

Article 35

Deposit and Regulation

1) The present Agreement shall be deposited with the Depositary who shall transmit certified true copies thereof to the Governments of all Contracting Parties which have signed the present Agreement or acceded to it.

2) As soon as the present Agreement enters into force, the text shall be transmitted by the Depositary to the Secretary-General of the United Nations for registration and publication, in accordance with Article 102 of the Charter of the United Nations.

12 International Plant Protection Convention (1951)

Preamble

The Contracting Governments, recognizing the usefulness of international co-operation in controlling pests and diseases of plants and plant products and in preventing their introduction and spread across national boundaries, and desiring to ensure close co-ordination of measures directed to these ends, have agreed as follows:

Article I

Purpose and Responsibility

1. With the purpose of securing common and effective action to prevent the introduction and spread of pests and diseases of plants and plant products and to promote measures for their control, the contracting Governments undertake to adopt the legislative, technical and administrative measures specified in this Convention and in supplementary agreements pursuant to Article III.

2. Each contracting Government shall assume responsibility for the fulfilment within its territories of all requirements under this Convention.

Article II

Scope

1. For the purposes of this Convention the term "plant" shall comprise living plants and parts thereof, including seeds in so far as the supervision of their importation under Article VI of the Convention or the issue of phytosanitary certificates in respect of them under Articles IV(1), (a) (iv) and V of this Convention may be deemed necessary by contracting Governments; and the term "plant products" shall comprise unmanufactured and milled material of plant origin, including seeds in so far as they are not included in the term "plants".

2. The provisions of this Convention may be deemed by contracting Governments to extend to storage places, containers, conveyances, packing material and accompanying media of all sorts including soil involved in the international transportation of plants and plant products

3. This Convention shall have particular reference to pests and diseases of importance to international trade.

Article III

Supplementary Agreements

1. Supplementary agreements applicable to specific regions, to specific pests and diseases, to specific plants and plant products, to specific methods of international transportation of plants and plant products, or otherwise supplementing the provisons of this Convention, may be proposed by the Food and Agriculture Organization of the United Nations (hereinafter referred to as "FAO") on the recommendation of contracting Government or on its own initiative, to meet special problems of plant protection which need particular attention or action.

2. Any such supplementary agreements shall come into force for each contracting Government after acceptance in accordance with the provisions of the FAO Constitution and Rules of Procedure.

Article IV

National Organization for Plant Protection

1. Each contracting Government shall make provision, as soon as possible and to the best of its ability, for

   a) an official plant protection organization, with the following main functions:

      (i) the inspection of growing plants, of areas under cultivation (including fields, plantations, nurseries, gardens and greenhouses) , and of plants and plant products in storage and in transportation particularly with the object of reporting the existence, outbreak and spread of plant diseases and pests and of controlling those pests and diseases;

      (ii) the inspection of consignments of plants and plant products moving in international traffic, and, as far as practicable, the inspection of consignments of other articles or commodities moving in international traffic under conditions where they may act incidentally as carriers of pests and diseases of plants and plant products, and the inspection and supervision of storage and transportation facilities of all kinds involved in international traffic whether of plants and plant products or of other commodities, particularly with the object of preventing the dissemination across national boundaries of pests and diseases of plants and plant products;

(iii) the disinfestation or disinfection of consignments of plants and plant products moving in international traffic, and their containers, storage places, or transportation facilities of all kinds employed;

(iv) the issue of certificates relating to phytosanitary conditions and origin of consignments of plants and plant products (hereinafter referred to as "phytosanitary certificates");

b) the distribution of information within the country regarding the pests and diseases of plants and plant products and the means of their prevention and control;

c) research and investigation in the field of plant protection.

2. Each contracting Government shall submit a description of the scope of its national organization for plant protection and of changes in such organization to the Director-General of FAO, who shall circulate such information to all contracting Governments.

Article V

Phytosanitary Certificates

1. Each contracting Government shall make arrangements for the issue of phytosanitary certificates to accord with the plant protection regulations of other contracting Governments and in conformity with the following provisions:

a) Inspection shall be carried out and certificates issued only by or under the authority of technically qualified and duly authorized officers and in such circumstances and with such knowledge and information available to those officers that the authorities of importing countries may accept such certificates with confidence as dependable documents.

b) Each certificate covering materials intended for planting or propagation shall be as worded in the Annex to this Convention and shall include such additional declarations as may be required by the importing country. The model certificate may also be used for other plants or plant products where appropriate and not inconsistent with the requirements of the importing country.

c) The certificates shall bear no alterations or erasures.

2. Each contracting Government undertakes not to require consignment of plants intended for planting or propagation imported into its territories to be accompanied by phytosanitary certificates inconsistent with the model set out in the Annex to this Convention.

Article VI

Requirements in Relation to Imports

1. With the aim of preventing the introduction of diseases and pests of plants into their territories , contracting Governments shall have full authority to regulate the entry of plants and plant products, and to this end, may:

a) prescribe restrictions or requirements concerning the importation of plants or plant products;

b) prohibit the importation of particular plants or plant products, or of particular consignments of plants or plant products;

c) inspect or detain particular consignments of plants or plant products;

d) treat, destroy or refuse entry to particular consignments of plants or plant products, or require such consignments to be treated or destroyed.

2. In order to minimize interference with international trade, each contracting Government undertakes to carry out the provisions referred to in paragraph 1 of this Article in conformity with the following:

a) Contracting Governments shall not, under this plant protection legislation, take any of the measures specified in paragraph 1 of this Article unless such measures are made necessary by phytosanitary considerations.

b) If a contracting Government prescribes any restrictions or requirements concerning the importation of plants and plant products into its territories, it shall publish the restrictions or requirements and communicate them immediately to the plant protection services of other contracting Governments and to FAO.

c) If a contracting Government prohibits, under the provisions of its plant protection legislation, the importation of any plant or plant products, it shall publish its decision with reasons and shall immediately inform the plant protection services of other contracting Governments and FAO.

d) If a contracting Government requires consignment of particular products to be imported only through specified points of entry, such points shall be so selected as not unnecessarily to impede international commerce. The contracting Government shall publish a list of such points of entry and communicate it to the plant protection services of other contracting Governments and to FAO, such restrictions on points of entry shall not be made unless the plants or plant products concerned are required to be accompanied by phytosanitary certificates or to be submitted to inspection or treatment.

e) Any inspection by the plant protection service of a contracting Government of consignments of plants offered for importation shall take place as promptly as possible with due regard to the perishability of the plants concerned. If any consignment is found not to conform to the requirements of the plant protection legislation of the importing country, the plant protection service of the exporting country shall be informed. If the consignment is destroyed, in whole or in part, and official report shall be forwarded immediately to the plant protection service of the exporting country.

f) Contracting Governments shall make provisions which, without endangering their own plant production, will reduce to a minimum the number of cases in which a phytosanitary certificate is required on the entry of plants or plant products not intended for planting , such as cereals, fruits, vegetables and cut flowers.

g) Contracting Governments may make provision for the importation for purposes of scientific research of plants and plant products and of specimens of plant pests and disease-causing organisms under conditions affording ample precaution against the risk of spreading plant diseases and pests.

3. The measures specified in this Article shall not be applied to goods in transit throughout the territories of contracting Governments unless such measures are necessary for the protection of their own plants.

Article VII

International Co-operation

The contracting Governments shall co-operate with one another to the fullest practicable extent in achieving the aims of this Convention, in particular as follows:

a) Each contracting Government agrees to co-operate with FAO in the establishment of a world reporting service on plant diseases and pests, making full use of the facilities and services of existing organizations for this purpose, and, when this is established, to furnish to FAO periodically the following information.

   (i) reports on the occurrence, outbreak and spread of economically important pests and diseases of plants and plant products which may be of immediate or potential danger;

   (ii) information on means found to be effective in controlling the pests and diseases of plants and plant products.

b) Each contracting Government shall, as far as is practicable, participate in any special campaigns for combating particular destructive pests or diseases which may seriously threaten crop production and need international action to meet the emergencies.

Article VIII

Regional Plant Protection Organization

1. The contracting Governments undertake to co-operate with one another in establishing regional plant protection organizations in appropriate areas.

2. The regional plant protection organizations shall function as the co-ordinating bodies in the areas covered and shall participate in various activities to achieve the objectives of this Convention.

Article IX

Settlement of Disputes

1. If there is any dispute regarding the interpretation or application of this Convention, or if a contracting Government considers that any action by another contracting Government is in conflict with the obligations of the latter under Articles V and VI of this Convention, especially regarding the basis of prohibiting or restricting the imports of plants or plant products coming from its territories, the Government or Governments concerned may request the Director-General of FAO to appoint a committee to consider the question in dispute.

2. The Director-General of FAO shall thereupon, after consultation with the Governments concerned, appoint a committee of experts which shall include representatives of those Governments. This committee shall consider the question in dispute, taking into account all documents and other forms of evidence submitted by the Governments concerned. This committee shall submit a report to the Director-General of FAO who shall transmit it to the Governments concerned, and to other contracting Governments.

3. The contracting Governments agree that the recommendations of such a committee, while not binding in character, will become the basis for renewed consideration by the Governments concerned of the matter out of which the disagreement arose.
4. The Governments concerned shall share equally the expenses of the experts.

Article X

Substitution of Prior Agreements

This Convention shall terminate and replace, between contracting Governments, the International Convention respecting measures to be taken against the Phylloxera vastarix of 3 November 1881, the additional Convention signed at Berne on 15 April 1889 and the International Convention for the Protection of Plants signed at Rome on 16 April 1929.

Article XI

Territorial Application

1. Any Government may at the time of ratification or adherence or at any time thereafter communicate to the Director-General of FAO a declaration that this Convention shall extend to all or any of the territories for the international relations of which it is responsible, and this Convention shall be applicable to all territories specified in the declaration as from the thirtieth day after the receipt of the declaration by the Director-General.
2. Any Government which has communicated to the Director-General of FAO a declaration in accordance with paragraph 1 of this Article may at any time communicate a further declaration modifying the scope of any former declaration or determinating the

application of the provisions of the present Convention in respect of any territory. Such modification or termination shall take effect as from the thirtieth day after the receipt of the declaration by the Director-General.

3. The Director-General of FAO shall inform all signatory and adhering Governments of any declaration received under this Article.

Article XII

Ratification and Adherence

1. This Convention shall be open for signature by all Governments until 1 May 1952 and shall be ratified at the earliest possible date. The instruments of ratification shall be deposited with the Director-General of FAO, who shall give notice of the date of deposit to each of the signatory Governments.

2. As soon as this Convention has come into force in accordance with Article XIV, it shall be open for adherence by non-signatory Governments, Adherence shall be effected by the deposit of an instrument of adherence with the Director-General of FAO, who shall notify all signatory and adhering Governments.

Article XIII

Amendment

1. Any proposal by a contracting Government for the amendment of this Convention shall be communicated to the Director-General of FAO.

2. Any proposal amendment of this Convention received by the Director-General of FAO from a contracting Government shall be presented to a regular or special session of the Conference of FAO for approval and , if the amendment involves

important technical changes or imposes additional obligations on the contracting Governments, it shall be considered by an advisory committee of specialists convened by FAO prior to the Conference.

3. Notice of any proposed amendment of this Convention shall be transmitted to the contracting Governments by the Director-General of FAO not later than the time when the agenda of the session of the Conference at which the matter is to be considered is dispatched.

4. Any such proposed amendment of this Convention shall require the approval of the Conference of FAO and shall come into force as from the thirtieth day after acceptacne by two-thirds of the contracting Governments.

5. The instruments of acceptance of amendments involving new obligations shall be deposited with the Diretor-General of FAO, who shall inform all contracting Governments of the receipt of acceptances and the entry into force of amendments.

Article XV

Denunciation

1. Any contracting Government may at any time give notice of denunciation of this Convention by notificaion addressed to the Director-General of FAO. The Director-General shall at once inform all signatory and adhering Governments.

2. Denunciation shall take effect one year form the date of receipt of the notification by the Director-General of FAO.

INDEX